Springer Series in Synergetics

Springer Series in Synergetics

Founding Editor: H. Haken

The Springer Series in Synergetics was founded by Herman Haken in 1977. Since then, the series has evolved into a substantial reference library for the quantitative, theoretical and methodological foundations of the science of complex systems.

Through many enduring classic texts, such as Haken's *Synergetics and Information and Self-Organization*, Gardiner's *Handbook of Stochastic Methods*, Risken's *The Fokker Planck-Equation* or Haake's *Quantum Signatures of Chaos*, the series has made, and continues to make, important contributions to shaping the foundations of the field.

The series publishes monographs and graduate-level textbooks of broad and general interest, with a pronounced emphasis on the physico-mathematical approach.

More information about this series at http://www.springer.com/series/712

Hermann Haken · Juval Portugali

Synergetic Cities: Information, Steady State and Phase Transition

Implications to Urban Scaling, Smart Cities and Planning

Springer

Hermann Haken
Center of Synergetics
University of Stuttgart
Stuttgart, Baden-Württemberg, Germany

Juval Portugali
Department of Geography and Human Environment
Tel Aviv University
Tel Aviv, Israel

ISSN 0172-7389 ISSN 2198-333X (electronic)
Springer Series in Synergetics
ISBN 978-3-030-63456-8 ISBN 978-3-030-63457-5 (eBook)
https://doi.org/10.1007/978-3-030-63457-5

This Springer imprint is published by the registered company Springer Nature Switzerland AG
The registered company address is: Gewerbestrasse 11, 6330 Cham, Switzerland

Preface

The last decades have witnessed the emergence of complexity theories of cities (CTC)—a domain of research that applies the various theories of complexity to the study of cities. As a consequence, there are now several theoretical perspectives on cities as complex systems; "synergetic cities" is one of them.

Synergetic cities (SC) is a theoretical perspective on cities that commences from the basic principles of synergetics. This is a field of research that deals with principles underlying the processes of self-organization in systems belonging to disciplines ranging from physics over biology and social sciences to the humanities including cognitive science.

The notion of SC that we present in this book has developed gradually out of the collaboration between the two of us. However, our aim in this book is not to summarize past studies but rather to extend the notion of SC and thus the domain of CTC. This, by adding to both three novel components: (1) the concepts of *Synergetic Inter-representation Networks* (SIRN), *Information Adaptation* (IA) and their conjunction SIRNIA—concepts that we developed in the past and elucidate the role of cognition in cities. (2) *Steady States* (StS) that are characteristic of the *longue durée* of cities, perhaps undergoing a smooth development. (3) *Phase Transitions* (PT) that are characterized by pronounced qualitative changes of important "indicators."

In particular, StS and PT are among the very basic aspects of complexity, which due to the specific development of the discourse on cities as complex systems, have not as yet received sufficient attention. The focus of interest in CTC was (still is) on the bottom-up emergence of new properties or structures in cities. But how/why does suddenly a new property emerges? What happens before and after its emergence? Such questions were largely overlooked in the CTC discourse. This is not the case with other domains (e.g., physics) in which one finds extensive discussions on StS and PT. And indeed, we start discussing PT in cities by means of analogies to case studies in physics, whereas we introduce new theoretical tools to study StS in cities. We then further elaborate the uniqueness of these processes in the dynamics of cities.

The book is conceived to reach a broad audience from a variety of fields starting with the young domain of CTC to more traditional disciplines such as human/urban geography, urban economy and sociology, city planning, architecture, urban design and more. Its backbone is a presentation of basic concepts without mathematics. But in line with the development of CTC, the concepts are complemented by mathematical approaches. Perhaps with one exception—a chapter on pattern recognition—all chapters are written in a pedagogical style to enable readers to familiarize themselves with the mathematical tools that are basic to synergetic systems theory, including stochastic processes (theory of chance events), information theory, Friston's Free Energy Principle, etc. All of them are useful tools for urbanism and are employed in our book. We hope that readers will enjoy reading it and find it useful for their own studies/research on the understanding of the marvelous phenomenon "city."

The book has greatly benefited from several of our studies in recent years. Thus, the second part of Chap. 3 is based on Sect. 4.2.2 of Portugali 2011. Chapter 13 largely follows our paper: Haken H, Portugali J (2019) "A synergetic perspective on urban scaling, urban regulatory focus and their interrelations." *Royal Society Open Science.* 6: 191087. http://dx.doi.org/10.1098/rsos. 191087. Chapter 14 is largely based on Haken H., Portugali J. (2017) "Smart cities: distributed intelligence or central planning?" In Pardalos P. M. and Rassia S. (Eds.) *Smart City Networks: Through the Internet of Things*, Springer, Heidelberg. Finally, Chap. 15 closely follows Portugali J. (2020) "Information adaptation as the link between cognitive planning and professional planning." In Gert de Roo, C. Yamu and C. Zuidema (Eds.) *Handbook on Planning and Complexity.* Edward Elgar Publishers.

We would like to close this preface with acknowledgments. In particular, Hermann Haken thanks his daughter Karin for her indefatigable help. Both authors thank Dr. Thomas Ditzinger and Mr. Gowrishankar Ayyasamy from Springer for their helpful cooperation. Special thanks also to Mrs. Shoshi Esronie from TAU for typing parts of the text.

Stuttgart, Germany
Tel Aviv, Israel
August 2020

Hermann Haken
Juval Portugali

Contents

Chapter 1
Introduction

1.1 Aim and Scope

Four concepts make the title of this book: *Synergetic cities* which is a view on cites as complex systems from the perspective of Haken's theory of synergetics; *information*, which is a view on cities as complex systems commencing from the perspective of information theory. Next come *steady state* and *phase transition* which are two central aspects of complex systems in general and of cities as complex systems. Our aim is to introduce and develop the above four notions and then to discuss their implication to three issues that stand at the core of current discourse on cities as complex systems: *urban allometery* (or *scaling*) and *smart cities*—both attract special attention in the discourse on cities of the last two decades, as part of the attempt to transform the study of cities into a science. The third issue, *city planning*, attempts to adapt the process of planning to the understanding, and reality, of cities as complex, adaptive self-organizing systems. Before we proceed, a few introductory words on the above four concepts and the rational of focusing on them is in place.

Synergetic cities. Complexity theory (or science) is an umbrella name for a set of theories that started to appear in the 1960s in order to deals with systems that are typified by the following properties: they are *open* in that they exchange matter, energy and information with their environment, they are in *far from equilibrium conditions* implying that they are never in rest, and yet they are more often than not highly ordered and tend to achieve their order spontaneously by means of *self-organization*; they exhibit long periods of *steady state* interrupted by short events of *chaos*, that often lead to *phase transition*, and much more. Synergetics is one of the founding theories in this new domain. All theories of complexity were applied to cities giving rise to *complexity theories of cities* (CTC)—a domain of research that studies cities as complex systems from the various perspectives of the "grand" complexity theories (Portugali 2011). And similarly to the core theories, each theory emphasizes a different aspect of the complexity of cities.

H. Haken and J. Portugali, *Synergetic Cities: Information, Steady State and Phase Transition*, Springer Series in Synergetics,
https://doi.org/10.1007/978-3-030-63457-5_1

Common to all CTC is that cities emerge from the bottom up. Synergetic cities too commences from this view but adds that once emerged, cities top—down determine ("enslave") the behavior of the urban agents and so on in circular causality. This view and property of synergetics makes it specifically attractive to study cities, as one of the central issues in the last 120 years of urban studies has always been the way cities affect their citizens: this issue attracted Simmel (1903) at the turn of the twentieth century with his classic *The Metropolisand Mental Life*, it is central to social theory oriented structuralist urban studies (e.g. Lefebvre 1970), and it is still discussed today in the context of urban allometry/scaling, regarding the relations between "Growth, innovation, scaling and the pace of life in cities" (Bettencourt et al. 2007).

Information. As shown below (Chap. 2), cities are *dual* and *hybrid* systems. Dual, in that each agent and the city as a whole is a complex system; hybrid, in that the city is composed of artifacts which are simple systems and urban agents which are complex systems. From these perspectives, central to the study of cities as complex systems is the interaction between the parts—the urban agents; as well as between the urban agents and the artifacts of which cities are composed (buildings, roads, … neighborhoods and whole cities). In these processes of interaction agents exchange information between themselves and between them and the information conveyed by the various urban artifacts and the city as a collective artifact. Our notions of *SIRN* (synergetic inter-representation networks), *IA* (information adaptation) and their conjunction—*SIRNIA*, were specifically designed to capture these processes.

Steady state and phase transition. Complex systems, cities included, are typified, firstly, by long periods of steady state during which, according to Synergetics, the system city is dominated by one or a few *order parameters* (OP). Secondly, by short events of strong fluctuations that interrupt the steady state and often entail a qualitative systemic change, that is, a *phase transition* (PT) that brings the system to a new steady state and so on. This is the "big picture", however; the more detailed one is that during steady state the system is interrupted by random fluctuations that the city (its OP) manages to "enslave"—only when such fluctuations happen in periods of instability, their effect might be dramatic and lead the system city to a PT and structural change. The fact is, that the general tendency in the domain of CTC is to focus on the bottom–up process of the emergence of a new systemic order and to ignore the dynamics of the long periods of steady state. As noted above, one of our aims is to correct this situation and elaborate on the dynamics of steady states and phase transition and the role of fluctuations in these processes.

As indicated by the title and as noted above, our aim is to discuss the implication of the above notions to three issues: *urban allometery* (or *scaling*), *smart cities* and *city planning*. While the three stand at the core of current CTC discourse, they differ considerably: Urban allometry studies are mathematically and ("big") data oriented representing the long history of attempts to develop a science of cities. Smart cities studies explore the implication of the so-called *4th industrial revolution* (Schwab 2016) with its smart devices—IoT, big data, AI, and the like—to various aspects of cities, ranging from transportation, through pollution and sustainability to governance and planning. Many studies in this domain are futuristic and somewhat utopic. Finally, similarly to, and in connection with, the study of cities, planning has a century long history of debates about the proper ways to plan cities. The appearance of CTC some three decades ago with notions such as self-organization and the like, gave rise to a study of the planning implications—theoretical and practical—of the finding that cities are complex adaptive systems, when the degree of adaptability may vary considerably between cities: While some cities flourish, others may have high debts, deteriorate, become obsolescent and lose their importance.

1.2 The Structure of the Book

The discussion in the book evolves in 15 chapters grouped into 3 parts. Part I lays the theoretical foundations. It starts by claiming that cities as complex systems differ from complex systems studied in the material and organic domains in that they are *hybrid complex systems* composed of artifacts which are simple systems and human urban agents which are complex systems; and, it is the urban agents that by means of their interaction, among themselves and with the urban artifacts, make the city a complex system (Chap. 2). Chapter 3 that follows is a short reminder of the theory of Synergetics, its basic concepts and language, and the early attempts to apply them to cities. Subsequent elaborations of the synergetic perspective on cities gave rise, firstly, to the notion of SIRN, then to the notion of IA that complements it and finally to their conjunction SIRNIA which is a theory that has a wider range of applications ranging from a solitary cognitive agent, through a sequence of agents to cities as hybrid complex systems. Chapter 4 thus introduces SIRNIA: first SIRN, next IA and finally their conjunction SIRNIA. The next two chapters introduce the two theoretical foundations of Synergetics: Chap. 5 the 1st Foundation which, as the title indicates, is a bottom–up approach, that is, from parts to order parameters, while Chap. 6 the 2nd Foundation which is a top–down approach—from sparse or big data to laws. Chapter 7 closes Part I by a novel discussion relating the Synergetics' 2nd Foundation and SIRNIA to Friston's *free energy principle* (FEP) regarding the tendency of complex biological systems (e.g. brain) to remain in *steady state* (StS).

Part II elaborates on two aspects/processes which are central to complex systems, yet for several reasons have not so far received sufficient attention in the domain of CTC: steady state and phase transition. In the context of synergetics, they are further associated with 'the slaving principle'—the process by which the system (the city) top–down effects the behavior of the parts (the urban agent), and the process of 'circular causality', that is, the continuous process of bottom–up/top–down mutual determination. Chapter 8 thus explores the process of steady state and the city, showing that a steady state entails, and is made possible by, a network of interdependencies of the actions of citizens.

Chapter 9 discusses several phenomena of phase transition in physics as a preparatory step toward Chap. 10 on phase transition and the city. Chapter 11 scrutinizes the slaving principle and circular causality in cities showing that in the urban realm it is often associated with a space-time diffusion process that in turn gives rise to variations in the nature of the urban order parameter(s).

Finally, Part III examines the implications to four issues that stand at the center of current CTC discourse: Urban allometery or scaling, the interplay between agents' motivation and the pace of life in cities, smart cities and urban planning. Thus, Chap. 12 shows that that the "classical" scaling law, that stands at the center of urban allometry, is typical of the steady-state periods of cities, while the strongly fluctuating phase transition periods, are typified by other scaling relationships. Chapter 13 further elaborates on the central claim of allometry that city size correlates positively with the pace of life in cities. Based on Ross and Portugali's (2018) finding that a city pace of life affects its citizens' motivational tendencies, Chap. 13 demonstrates how this finding further affects the dynamics of cities.

Chapter 14 turns attention to the implications of our findings to the smartification of cities due to the introduction of smart *Information communication technologies* (ICTs). In particular, Chap. 14 deals with information production by humans and smart devices and their interplay. It shows that the latter enhances the intellectual load of humans, a counterintuitive result. The challenge facing smart cities is thus, to identify a steady state that maximizes the relative advantage of the human sensorium and intelligence and that of the artificial ones. Last but not least, Chap. 15 focuses on the implications to urban planning. It, firstly, introduces *cognitive planning* as a basic cognitive capability of every urban agent, that is active in cities in addition to the regular *professional planning*. Then, looking at these two kind of planning from the perspective of the theoretical perspective of our book, it is shown how the two are interwoven with each other in a kind of circular causality.

The book concludes with a few preliminary comments on the COVID-19 pandemic which, as already can be seen, exhibits many of the ingredients of a complex system as well as strong links to cities, while its possible future effects on cities and their dynamics have yet to be seen.

References

Bettencourt, L. M. A., Lobo, J., Helbing, D., Kühnert, C., & West, G. B. (2007). Growth, innovation, scaling, and the pace of life in cities. *Proceedings of National Academy Science U.S.A., 104*(17), 7301–7306. https://doi.org/10.1073/pnas.0610172104.

Portugali, J. (2011). *Complexity, cognition and the city*. Berlin/Heidelberg/New York: Springer.

Ross, G. M., & Portugali, J. (2018). *Urban regulatory focus: A new concept linking city size to human behavior*.

Schwab, K. (2016). *The fourth industrial revolution* (Kindle). Switzerland: World Economic Forum.

Simmel, G. (1903). "The Metropolis and Mental Life", trans. by H. H. Gerth and C. Wright Mills), in Kurt H. Wolff, The Sociology of Georg Simmel (New York, 1964), 424

Lefebvre, H. (1970). La révolution urbaine Paris: Gallimard, Collection "Idées"

Part I
Theoretical Foundations

Chapter 2
Cities as Hybrid Complex Systems

2.1 Cities as Complex Systems

The last decades have witnessed two interrelated developments: Firstly, the emergence of *complexity theories* (CT) and secondly the emergence of *complexity theories of cities* (CTC). CT is a set of theories that started to appear in the 1960s, referring to open and complex systems that exhibit phenomena of selforganization, chaos etc. With the exception of Humberto Maturana's *Autopoiesis theory* (Maturana and Varela 1973), the founding complexity theories were first developed in the material sciences (physics, chemistry, etc.) closely linked to thermodynamics and with reference to phenomena such as fluid dynamics, laser a.s.o. At subsequent stages complexity theories were applied to other domains; in particular, their application to the organic domains entailed the inclusion in the vocabulary of complexity theories the notion of *adaptation* as a basic property of complex systems: Originally a life science property—'adaptation' was added to complexity theory by Nobel Laureate in physics Gell-Mann (1994, p. 17), who in his *The Quark and the Jaguar* coined the notion of CAS—*complex adaptive systems*; at a subsequent stage, computer scientist Holland (2006), further developed and popularized it.

All complexity theories were very successfully applied to cities, entailing the second development: the emergence of CTC (complexity theories of cities)—a domain of research that applies the various theories of complexity to the study of cities. And following the core theories, here too, the first wave of applications defined cities as complex systems in a physicalist sense, while the second wave treated cities as complex adaptive systems. CTC studies have shown that cities are open, complex, dynamical systems largely based on *selforganization,* associated with information processing, society and population dynamics and so on.

The above successful applications were a result of the fact that cities as complex systems share many similarities with natural material and organic complex systems. For example, similarly to material complex systems cities are open, complex, far from equilibrium, typified by self-organization, chaotic events, fractal structure and

H. Haken and J. Portugali, *Synergetic Cities: Information, Steady State and Phase Transition*, Springer Series in Synergetics,
https://doi.org/10.1007/978-3-030-63457-5_2

the like, and similarly to organic complex systems they are adaptive systems capable of adapting their structure to changing internal and external conditions, and more. On the other hand, however, cities also differ from natural complex systems in that they are *hybrid* and as a consequence *dual complex systems*.

2.2 Cities as Hybrid Complex Systems

Cities are composed of material components and organic/human components. As a set of material components alone (buildings, roads, bridges, ... whole cities and metropoles), the city is an artifact; the artifacts of which a city is composed and the city as an artifact, lack some of the basic properties of complexity: they are incapable of interacting among themselves and with the environment and as such the city is a simple system. On the other hand, as a set of human components—the urban agents (individuals, families, households ...)—the city is a complex system. The city is thus a *hybrid simple-complex system* and it is the urban agents that by means of their interaction—among themselves, with the city's material components and with the environment—transform the artifact city into the complex-artificial system 'city'.

As a complex artificial system, the city emerges out of the interactional activities of its agents that build the artifacts and thus the city, that once it emerges, it affects ("enslaves", in the language of synergetics) the behavior of its agents and so on in circular causality—a process that in the domain of social theory is termed *socio-spatial reproduction* (e.g. Lefebvre (1974); Giddens (1984)). In other words, the city is a large-scale collective and complex artifact that, on the one hand, due to its inhabitants and users (the urban agents), interacts with its environment, while on the other, due to its size, functions as an environment for the large number of people that live and act in cities. This latter effect and property of cities is becoming more and more prominent in society as the proportion of people living in cities is growing—as is well recorded (Wimberley et al. 2007), the last century has witnessed the fastest population growth in human history and the fastest urbanization processes with the result that, for the first time, more than 50% of the world population lives in cities. The city in this respect is a *complex artificial environment* (Portugali 2011, Chap. 11).

2.2.1 Cities as Dually Complex Systems

As discussed in some length in CCCity (*Complexity, Cognition and the City*, Portugali 2011), the city as a whole is a (hybrid) complex system and each of its parts (agents) is also a complex system. This is not the case with material complex systems in which complexity is a property of the global systems but not of the parts. Organic complex system are different as each of their parts is a complex system too. In an organic complex system, each of the parts and the system as a whole, are typified by the property of adaptation which as we've seen above, gave rise to the notion of

CAS. Cities too are CASs, but their hybridity differentiates them from natural CAS. To see how, we have to appreciate the links between artifacts, culture and evolution.

Artifacts such as ideas, habits, texts, buildings, roads and whole cities, are the stuff of *cultural evolution*—the process by which the various artificial objects that form culture are learned and transmitted between individuals and collectivities and in the process change over time. As we elaborate in Chap. 11, based on the neo-Darwinian synthesis of genetics and evolution, the theories of cultural evolution (Cavalli-Sforza and Feldman 1981; Creanza et al. 2017) suggest a process reminiscence of Darwinian evolution, but fundamentally differs from it with respect to time-scale: compared to Darwinian evolution, cultural evolution is extremely fast. One reason for that is implicit in Cavalli-Sforza and Feldman's (ibid.) theory: Biological evolution is based on genetic transmission, while cultural evolution and transmission on *learning*, which similarly to genetic transmission embodies the potential for a leaning mistake, that implies a *cultural mutation*. However, unlike biological transmission, which occurs between parent and child of successive generations only, cultural transmission occurs between parent and child of successive generations, between neighbors, and between neighbors of two successive generations. These are termed by Cavalli-Sforza and Feldman (ibid.), respectively, *vertical*, *horizontal* and *oblique* transmissions. Note that horizontal and oblique transmissions imply the space-time *diffusion* of the cultural mutations. But see further discussion in Chap. 11.

The aforementioned time-scale difference implies that since the parts of organic complex systems (e.g. plants or animals) are subject to the slow process of natural Darwinian evolution, the short-term feedback effect of the global system (e.g. a flock of birds or fish) on the parts (e.g. a single fish or bird) is negligible and thus the property of duality of such systems can be ignored. The situation is different with respect to the hybrid complex systems 'cities', as their agents are simultaneously subject to two evolutionary processes: The very slow natural evolution and the very fast cultural evolution whose effect on the urban agents is instantaneous—urban agents have thus to adapt to the fast changing urban environment. How do they adapt to fast cultural changes? By means of their cognitive capabilities! The implication is that we have to engage the cognitive capabilities of the urban agents in the study of the dynamics of cities.

2.2.2 *Complexity, Cognition and the City*

CCCity (ibid.) was a first attempt in this direction with emphasis on the capability of *cognitive mapping*. Namely, that similarly to other animals, the behavior of the complex parts of the city (the urban agents) is mediated, and thus strongly influenced, by their cognitive maps of the city. This is significant, because studies on "systematic distortions in cognitive maps" of humans (Tversky 1992; Portugali 2011, Chap. 6) have shown that "the map is not the territory", namely, that cognitive maps are not one-to-one representations of the environment; rather, they are often systematically distorted in several specific ways.

In a subsequent study Portugali (2016) directed attention to a second cognitive capability by means of which urban agents' adapt to fast cultural changes, namely, their relation to *time* and by implication to planning and design. The urban agents—the parts of the complex system city—are parts of a special kind: they are typified by *chronesthesia* that is, the ability to mentally travel in time; back to the past and forward to the future. Unlike cognitive mapping, the cognitive property of chronesthesia seems to be unique to humans (Suddendorf and Corballis 2007). One example of mental time travel (MTT) is *prospective memory* (Haken and Portugali 2005), referring to human ability to remember to perform an intended or planned action.

A third capability concerns planning as studied in the research domain of *cognitive planning* (Miller et al. 1960; Ormerod 2005). Humans are, in this respect, natural planners and designers. However, not only do humans have this ability to mentally travel in time, but they also cannot avoid mentally travel in time (Portugali 2016); studies show that "unlike other animals," human beings spend about half of their waking hours "thinking about what is not going on around them, contemplating events that happened in the past, might happen in the future, or will never happen at all" (Killingsworth and Gilbert 2010). So much so, that "stimulus-independent thought" or "mind wandering" has been shown to be the brain's default mode of operation (Raichle et al. 2001; Buckner et al. 2008). It means that much of urban agents' behavior is determined not by response to the present situation in the city, but in response to an urban reality that doesn't yet exist and might never exist.

To conclude, artifacts are the product of human interaction and as noted above, artifacts are simple systems and as such cannot interact—their interrelations in the context of urban dynamics is mediated by the urban humans agents.[1] Yet it must be emphasized that artifacts are not just the outcome of human interaction; rather they are also *the media of interaction* between the urban agents: Artifacts such as texts, cities, buildings or roads are *external representations* of ideas, intentions, memories and thoughts that originate and reside in the mind/brain of urban agents; that is to say, of *internal representations*. However, exactly as material artifacts (e.g. buildings, roads, …) cannot directly interact among themselves, so also ideas, thoughts, intentions, plans and other internal representations cannot. They interact by means of the externally represented artifacts; be they utterances, texts, clothes … buildings, neighborhoods and whole cities and metropoles. Urban dynamics thus involves an ongoing interaction between external and internal representations.

The notions of SIRN (Synergetic inter-representation networks), IA (information adaptation) and their conjunction—SIRNIA introduced in Chap. 4 below, are specifically designed to capture the dynamics of cities as hybrid complex systems and as dually complex systems.

[1]Having said this, and as elaborated in Chap. 14 below, with the introduction *of internet of things* (IoT), smartification, computer networks and the like, this situation is changing and soon we might have "artificial urban agents".

2.3 On Methodology: How to Study Cities as Hybrid Complex Systems?

2.3.1 The Two Cultures of Cities

Complexity theory is a relatively young research domain as just noted, while the modern study of cities has a long history that started in mid nineteenth century with von Thünen's (1826/1966)*Isolated State*. The implication: CTC did not arrive to a 'terra incognita', but rather to an evolving "urban terra" into which it had and still have to integrate. Looking at the study ("terra") of cities prior to the emergence of the CTC domain, one can identify a pendulum moving between *two cultures of cities*—two methodological approaches that echoes Snow's (1964) observation regarding the gap between *The Two Cultures* of science as a whole: A quantitative-analytic approach versus social-theoretic hermeneutic one, with a gap of miscommunication between the two. In the quantitative-analytic stream, one can further observe inductive versus deductive approaches. For example, with respect to systems of cities, between the inductive *rank-size* approach pioneered by Auerbach (1913), versus the deductive *Central Place Theory* of Christaller (1933). Somewhat similarly, in the social theoretic hermeneutic domain one can identify, e.g., structuralist studies that commence from "deep" social structure (similar to top–down), versus phenomenological studies that emphasize the subjectivity of individuals and places (similar to bottom up).

And what about complexity theory and CTCs? What position or side they take in face of the above dichotomies? To which culture of cities they are inclined? On the face of it, CT is a quantitative-analytic approach that originated in the sciences—whose basic concepts (e.g. PT, StS, emergence, entropy) are taken from physics. Accordingly, CT arrived with a new conceptual framework and most importantly with new quantitative, statistical and mathematical formalisms. In fact, many of CTCs early, as well as current, practitioners were and are physicists applying their conceptual and mathematical tool-kits to the domain of cities. It is therefore of no surprise that most CTC practitioners tend to perceive themselves as advancing a new and more sophisticated version of the "old" quantitative analytic approach to cities.

On the other hand, however, if you examine some of the basic concepts of the various complexity theories, you find that they have found in matter properties hitherto attributed, and similar, to properties and concepts typical of the hermeneutic domains (Portugali 1985). For example, in 'classical' physics' predictability and by implication certainty and causality, were traditionally seen as basic properties and capabilities of the quantitative-analytic-mechanistic approaches, while their negations—un-predictability, uncertainty and non-causality were seen as a consequence of insufficient data. Classical physics had (still has) a strong influence of the quantitative approach to cities, as well as on the social sciences in general [e.g. August Comte (1798–1857), one of the founders of sociology, has suggested calling this science *social physics*]. In the "soft" hermeneutic-humanistic domains, per contra (e.g. history), uncertainty and 'un-predictability' were always seen as basic properties of human (historical, political, …) processes. Scientific developments such as quantum physics (e.g. Heisenberg's uncertainty relation) and Chaos theory (e.g. the 'butterfly effect') have shown that this does no more apply to 20th century physics. The same applies to the various complexity theories—they refer to natural material and organic systems which are un-predictable, un-certain and not causal; not because of poor data, but ontologically so.

In particular, however, the above similarity between complexity and hermeneutic-humanistic views is prominent with respect to two basic properties of complex systems that form the focus of the present book—StS and PT: complex systems are typified by long periods of StS interrupted by short periods of strong fluctuations that often result with PT and the emergence of a new StS. To the latter, the theory of synergetics adds that following a phase transition and the emergence of new structure and order (parameter), the new order "enslaves" the parts of the system and so on in circular causality. Similar perceptions typify social theory views: social evolution is characterized by long periods—termed "*epochs*", "*age*" *periods*, … of relative stability when a given regime dominates, followed by periods of chaos and drastic change—*revolutions* in the parlance of social theory; and following a revolution, once again a relatively stable period prevails by means of a process of *socio-spatial reproduction*—identical to the slaving principle and circular causality of synergetics. (See further discussion on this in Chap. 11). From this perspective CTCs are akin to the hermeneutic approaches of cities, specifically so since cities, as noted above, are hybrid complex systems.

The fact is, however, that compared to the other CTCs' notions such as emergence, bottom–up or self-organization, StS and PT have received only marginal attention and when they are being referred to in CTC, they are presented in the most general way. The reason: there is an implicit link between methodology and content—between the methodology you favor and the research topics you choose: researchers in the "hard" quantitative-analytic "conviction" are specifically attracted by data rich research topics, while researchers in the "soft" hermeneutic camp by qualitative research issues—social, ethical, political, cultural and so on. Thus, in the context of urban studies, there has developed a kind of informal, implicit division of labor where most CTCs studies focus on data-rich, short-term, somewhat anachronistic more technical topics that afford quantification, leaving to students of the social theory oriented hermeneutic approaches, the qualitative long-term aspects of urban dynamics (to which StS and PT processes belong)—to our view, the really burning issues that stand at the core of twenty-first century problematics of cities. Currently, this division became even sharper due to the availability of big data, IoT and the like that according to Anderson (2008) indicates "The end of theory", since "The Data Deluge Makes the Scientific Method Obsolete". Whether CTC practitioners accept this view or not, the fact is that CTC studies on urban dynamics are strongly influenced by the "the data deluge".

Is the methodological gap between the two cultures of cities inevitable? Our view is that this is not the case. Firstly, as already suggested, due to the above noted similarity between CT and hermeneutic domains there is a potential that "Complexity theory [can function] as a link between space and place" (Portugali 2006, 2011), that is, between the two cultures of cities. The theory of synergetics and its extension "synergetic cities", with a focus on SIRNIA, StS and PT that we develop below, is part of an attempt to pave the way toward the realization of the above potential.

2.3.2 *Small Data is Beautiful*

How to study cities as complex systems? The fashionable answer these days, specifically in the context of smart cities, is *Big Data*: The new ICT with their data deluge, data mining and visual analytic techniques, make it possible for the first time to make use of the huge amount of data that is accumulating in the internet and other networks. Yet, the title of this section suggests an alternative view—it rephrases the title of Schumacher's (1973) book *Small is beautiful: A Study of Economics As If People Mattered.* This book was a critique of mid twentieth century's economic "gigantism" in the form of mass production, mass media, fast economic growth and urbanization that according to Schumacher entailed the de-humanization of people and their cultures. The subtitle of this section should thus implied: *A Study of cities as if people mattered.*

The common view in CTC is that cities emerge from the bottom–up out of the interactive behavior between their parts, the urban agents. It is also common to follow economics' "homo economicus" and to assume that urban agents behave and act in a consistently rational self-interested way. Our view, as implied from the aforementioned view of cities as dually complex systems, is that urban agents behave and act according to their subjective interpretation of the information conveyed by the city and the elements of which it is composed—in a way specified by our SIRNIA model (Chap. 4). The data about such behavior and action is generated by cognitive and brain sciences and is termed here *small data.* We use it here to understand and model the interactive behavior and action of the urban agents among themselves and with their urban environment.

According to our 'synergetic cities' view, and similarly to other CTC, cities as complex systems emerge from the bottom–up, that is, by means of the interaction between their parts—the urban agents (inhabitants, users, …). However, unlike other CTCs, synergetics suggests that once they come into being, cities top–down determine ("enslave") the action and behavior of their parts—the urban agents, and so on in circular causality. Accordingly, for the first, bottom–up, part of the urban process 'small data' is needed. Big data refers to the global structure and properties of a city as a whole and is thus needed to understand, study and model the top–down processes by which the city with it OP (e.g. urban scaling) effects the action and behavior of the urban agents (see discussion in Chap. 6). Big and small data thus do not negate each other, but rather co-exist in complementary relations. These complementary relations are at the core of the two foundations of the theory of synergetic: The 1st Foundation that commences from the bottom–up and continues top–down (Haken 1977), and The 2nd Foundation of synergetics (Haken 2006) that commences from the top–down taking into consideration the insight from the first foundation. The discussion in this book thus evolves as a back and forth movement and discourse between small and big data views: Chap. 5 develops a 'bottom–up' formalism—from parts to whole (OP), while Chap. 6 starts with big data from the top–down; this movement and discourse continues in the subsequent chapters of the book.

References

Auerbach, F. (1913). Das Gesetz der Bevoelkerungskonzentration. *Petermanns Geograph Mittl, 59,* 74–76.

Buckner, R., Andrews-Hanna, J., & Schacter, D. (2008). The brain's default network: Anatomy, function and relevance to disease. *Annals NY Academic Science, 1124,* 1–38.

Cavalli-Sforza, & Feldman (1981). *Cultural transmission and evolution: A quantitative approach.* Princton NJ: Princton Univ Press.

Chris, A. (2008). *The end of theory: The data deluge makes the scientific method obsolete.*

Christaller, W. (1933/1966). *Central Places in Southern Germany.* Englewood Cliffs, NJ: Prentice Hall

Creanza, N., Kolodny, O., Feldman, M. W. (2017) Cultural evolutionary theory: How culture evolves and why it matters. *PNAS, 114*(30), 7782–7789.

Gell-Mann, M. (1994). *The Quark and the Jaguar: Adventures in the simple and the complex.* New York: Freeman.

Giddens A. (1984). The Constitution of Society: Outline of the Theory of Structuration. Cambridge: Polity

Haken, H. (1977). *Synergetics—An introduction.* Berlin/Heidelberg/New York: Springer.

Haken, H., & Portugali, J. (2005). A synergetic interpretation of cue-dependent prospective memory. *Cognitive Process, 6,* 87–97.

Haken H. (2006). Information and Self-Organization: A Macroscopic Approach to Complex Systems, (3rd enlarged ed.). Springer

Holland, J. H. (2006). Studying complex systems. *Journal of Systems Science and Complexity, 19*(1), 1–8.

Killingsworth, D. T., & Gilbert, A. (2010). Wandering mind is an unhappy mind. *Science, 330*(6006), 932.

Lefebvre H (1974). La Production de l'espace. Paris: Anthropos. English translation, 1995. The Production of Space. Oxford: Blackwell.

Maturana, H., Varela, F. (1973). Autopoiesis: The organization of the living, a 1973 paper reprinted in: Autopoiesis and Cognition (Maturana and Varela, 1980), pp. 63–134.

Miller, G. A., Galanter, E. H., & Pribram, K. H. (1960). *Plans and the structure of behavior.* New York: Holt Rinehart & Winston.

Ormerod, T. C. (2005). Planning and ill-defined problems. In R. Morris & G. Ward (Eds.), *The cognitive psychology of planning* (pp. 53–70). Hove: Psychology Press.

Portugali, J. (2006). Complexity theory as a link between space and place. *Environment and Planning A, 38,* 647–664.

Portugali J (1985). Parallel currents in the natural and social Sciences. In: Portugali J (ed) Links Between Natural and Social Sciences. A special theme issue of Geoforum 16(2): 227– 238.

Portugali, J. (2011). *Complexity, cognition and the city,* Berlin/Heidelberg/New York: Springer.

Portugali, J. (2016). *Interview in Lisa Kremer: What's the Buzz about smart cities?* Tel Aviv University.

Raichle, M. E., MacLeod, A. M., Snyder, A. Z., Powers, W. J., Gusnard, D. A., & Shulman, G. L. (2001). A default mode of brain function. *Proceedings of National Academy Science USA, 98*(2), 676–682.

Schumacher's (1973). *Small is beautiful: A study of economics as if people mattered.* London: Blond & Briggs Ltd.
Snow CP (1964) The Two Cultures and a Second Look. Cambridge: Cambridge Univ Press
Suddendorf, T., & Corballis, M. C. (2007). The evolution of foresight: What is mental time travel, and is it unique to humans? *Behavioral and Brain Science, 30*(3), 299–313, 315–351.
Thünen JH von (1826/1966) von Thünen Isolated State. An English translation (P Hall, ed) Oxford: Pergamon
Tversky, B. (1992). Distortions in cognitive maps. *Geoforum, 23,* 131–138.
Wimberley, R., Morris, L., & Fulkerson, G. (2007). Mayday 23: World population becomes more urban than rural. *Rural Sociologist, 27,* 42–43.

Chapter 3
Synergetics: A Short Reminder

3.1 Synergetics

Synergetics ("science of cooperation") is an interdisciplinary field of research that aims at unearthing general principles underlying selforganization in open complex systems. These systems are composed of many individuals, parts, elements that interact/communicate with each other, and they exchange energy, matter, and/or information with their surroundings. And, most importantly, they are capable of producing spatial, temporal or functional structures without any external "organizer" ("selforganization"). The four basic concepts of Synergetics are:

(1) Order parameters (OPs)
(2) Slaving principle
(3) Circular causality
(4) Control parameter.

Order parameters play a double role (Fig. 3.1): They describe a macroscopic property of the considered system, and they prescribe the behavior of its individual parts via the slaving principle. In turn, the parts generate the OPs by means of their collective behavior ("circular causality"). A simple example may illuminate the concepts. The language of a nation is an OP. It characterizes a macroscopic property of the people, and it prescribes ("enslaves") their behavior. When a baby is born s/he learns this specific language and later, carries it further. In this way the individuals keep this very language alive, i.e. circular causality. Clearly, a language can "live" only, when its speakers live. Thus, an obvious control parameter is the amount of food. Below a critical value, the population dies out. This example is by no means far-fetched: the "function" of all synergetic systems requires a sufficient amount of energy supply. Our example reveals the role of "time scale separation". The life span of an order parameter (the specific language) is much larger than that of the action of the individuals. Actually, in the general frame of Synergetics it doesn't

H. Haken and J. Portugali, *Synergetic Cities: Information, Steady State and Phase Transition*, Springer Series in Synergetics,
https://doi.org/10.1007/978-3-030-63457-5_3

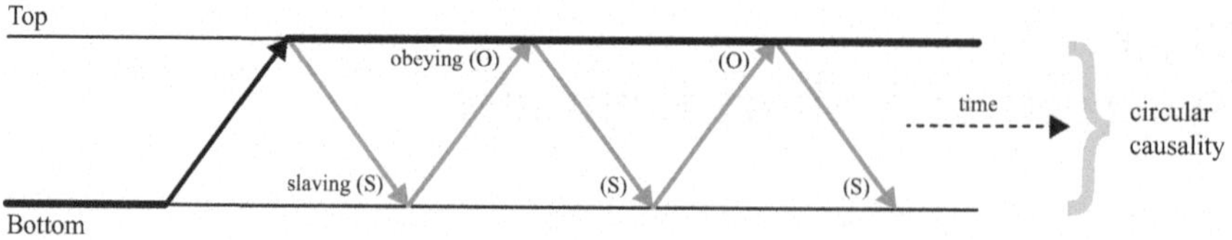

Fig. 3.1 The local interaction/synergy between parts (bottom) gives rise to an order parameter (top) that then enslaves the behavior of the parts (bottom). By 'obeying', the parts strengthen and reproduce the order parameter and so on in a so-called circularly causal fashion

matter, whether *new* individuals participate in the process, or whether the actions of the same individuals are reactivated time and again. Obviously, the life-time of the control parameter ("food supply") must be the largest. In many cases we will treat it as a given constant.

The concept of OPs is a powerful means for information "deflation" (a denotation coined in our IA-book—see below Chap. 4). Instead of describing the actions of each individual (part) we may refer to a collective property. On this macroscopic level we may deal with possible interplays between OPs. E.g. languages may compete with each other. An example is provided by the USA, where first English and German were spoken, but then following a vote, English has "won". Switzerland provides us with an example of the coexistence of several languages etc. What does the language example tell us about the origin of language as OP? Nothing at all!

To elucidate the coming into existence of OPs ("emergence") and the role of control parameters we present another simple example. Let us consider the behavior of swimmers in a circular swimming pool. If the density of the swimmers is low, the individuals may swim as they wish. But when the density rises beyond some critical value, the swimmers impede each other quite heavily—unless they start swimming in clockwise or anticlockwise circles. This may happen without any orders of a "life guard" (i.e. by selforganization). We may argue that first only few swimmers take the initiative to swim in a specific direction to induce others to follow until the fully ordered state is reached. Thus a transition from the formerly disordered "chaotic" state to a highly ordered one (having a "structure") has happened. In analogy to disorder-to-order transitions (or vice versa) in physics (melting of ice into water, onset of ferromagnetism) we call such transitions "phase transitions". Our swimming pool example exhibits a further widespread phenomenon of phase-transitions: spontaneous *symmetry breaking*. While "in principle", both clock- and anticlockwise motions are feasible ("symmetry of the solution") only one can be realized ("symmetry breaking"). Which one is chosen, depends on the spontaneous initiative of few people, or on convention. In the context of Synergetics spontaneous actions of one or few parts/participants are treated as chance events that are termed "fluctuations".

Furthermore, our example provides us with the insight that fluctuations can play a decisive role *only* when the control parameter is close to a "critical value" (here: critical density). At higher densities, additional swimmers must follow the established OP, or in other words, the OP is stable against fluctuations. Only the simultaneous collective action of all swimmers can change this "situation". Actually the same "mechanism" makes a language stable against various kinds of fluctuations, e.g. spelling errors, wrong wording, use of foreign language by only few people.

3.2 Synergetic Cities

The above examples of language and swimming as means to convey the basic principles of Synergetics, are intimately related to cities as complex synergetic systems. With respect to language, it is interesting to note that the links between languages and cities have a long history. While a full scale study of these links has yet to be done, and is thus beyond the scope of our book, a few notes/landmarks might be proper in the present context: In 1913 German physicist Felix Auerbach published a paper showing that the rank-size distribution of cities is typified by a power law distribution. Some three decades later American linguist and philologist Zipf (1949) demonstrated that the power law distribution characterizes also the frequency distribution of words in a language as well as the size distribution of elements in several other systems. While Auerbach's "Law of population concentration" has largely been forgotten, Zipf's study has left strong impression resulting in the fact that the power law is commonly called *Zipf Law,* while its appearance in the study of cities is often being referred to as "Zipf's law of cities". Actually, this law has all the features of an OP: it describes and prescribes a specific macroscopic structure that is brought about by self-organization in accordance with circular causality. More recently, the power law distribution of a network, including an urban network, was shown to be a result of, and indication to, the complexity of that network (Barabási 2016; Batty 2013). Finally, the starting point for the allometric cities that we discuss below (Chap. 12), is, in fact, the 1913 "Law of population concentration" by Auerbach. In the context of urban allometry, the role of the power law of cities as OP is even more pronounced as it describes and prescribes also the motivational and behavioral tendencies of urban agents that give rise to the city OP that in its turn affects ("enslaves") their motivation and behavior and so on in circular causality.

"Our language" writes Wittgenstein (1953) in his *Philosophical Investigations,* "can be seen as an ancient city: a maze of little streets and squares, of old and new houses, and of houses with additions from various periods; and this surrounded by a multitude of new boroughs with straight regular streets and uniform houses." Wittgenstein is using cities metaphorically in order to develop his view on language. Architects Alexander and Hillier developed theories suggesting that the morphology of a city is literally a language of patterns (Alexander et al. 1977) that in analogy to spoken language, has a (space) syntax (Hillier 2016), semantic and pragmatics. As a consequence, with minor modifications our above text on language can be directly applied to cities.

Similarly, to a language, a city can be seen as an OP that came into being out of the interaction between the people/citizens but once emerged, it characterizes the macroscopic property of the people, and prescribes ("enslaves") their behavior.

When a baby is born s/he learns the specific urban forms, rules and phenomena and later, carries it further. In this way the individuals keep this very urban language alive, i.e. circular causality. Clearly, a city can "live" only, when its citizens live and its material structures (artifacts) are maintained. Thus, obvious control parameters are the amount of food and material needed to sustain the city's inhabitants and artifacts. Below critical values, the population dies out and the material components of the city disintegrate (increase of entropy). Our example again reveals the role of "time scale separation". The life span of an order parameter (e.g. the specific city) is much larger than that of the action of the individuals. Obviously the life-time of the control parameter ("food and/or material supply") must be the largest. In many cases we will treat it as a given constant.

Swimming is a specific example for a general property of Synergetics, namely *movement*. Thus, to elucidate the coming into existence of OPs (emergence) in a city, and the role of control parameters we may refer to movement behavior in a city—of pedestrians on pavements or car drivers on the roads. If the density of the pedestrians or cars is low, the individuals may move (within limits) as they wish. But when the density rises beyond some critical value, the pedestrians or drivers in the cars impede each other quite heavily—unless they start synchronize their movement. This often happen spontaneously, i.e. by selforganization. We may argue that first only few pedestrians or drivers take the initiative to move in a specific pace to induce others to follow until the fully ordered state is reached. Thus a transition from the formally disordered "chaotic" state to a highly ordered one (having a "structure") has happened. We call such transitions "phase transitions".

Traffic on highways is typified by phase transitions; quite often even several different phases may arise: uncorrelated movements, a homogeneous flux, density waves (including stop and go), traffic jam (standing still). In the cities of India it is very common to see pedestrians, bicycles, rickshaws, cars, all in the same street and self-organizing their movement; similar scenarios have been recorded in several other countries including Ethiopia and Vietnam. The controversial notion of *shared space*, originally suggested by Dutch traffic engineer and designer Hans Monderman, claims that stripping roads of traffic lights and other instruments that control pedestrians' and cars' movement, will make our roads much safer and efficient as pedestrians and drives will become more conscious of each other and thus coordinate their movement (Karndacharuk et al. 2014). As can be seen, implicit in this claim is the view that bottom-up self-organization is more efficient than top-down controlled organization. This approach was already applied in several cities around the world (e.g. Bern, Switzerland). In fact, to a large extent we are approaching a shared space situation with the current introduction of electric two-wheel vehicles of various kinds (e.g. scooters).

3.3 The Four Paradigms of Synergetics

While so far we have illuminated the basic concepts of synergetics by means of examples of human behavior/cognition, it must be noted that they have originated and thus also apply to the inanimate world. Actually, the analysis of the function of the light source LASER (light amplification by stimulated emission of radiation) has laid the basis of the development of Synergetics and as such was the first of several case studies that became paradigmatic case studies and a convenient way to convey the principles of synergetics. They are: the laser paradigm, the pattern formation paradigm, the pattern recognition paradigm and the finger movement paradigm. They are described next.

3.3.1 The Laser Paradigm

The generic paradigm of synergetics is the process that produces the phenomenon of *laser* as described in Fig. 3.2. A typical example of the instrument laser is of a glass tube, with two mirrors at its ends, filled with a gas of atoms or molecules. The mirrors reflect the light that runs in axial direction causing the light waves to strongly interact with the individual atoms. One of the mirrors is semi-transparent so that the laser light can eventually emerge through it. The atoms/molecules are excited by an electric current sent through the tube. Unlike a regular lamp in which the excited individual atoms emit individual independent light waves, in the laser, a typical act of self-organization occurs: the individual electrons in the atoms correlate their movements and generate a beautifully ordered coherent light wave (Figs. 3.2 and 3.3). Nobody tells the laser system how to behave in such a coherent fashion; it finds its well-ordered behavior by itself.

In Chaps. 5 and 6 we will show how our qualitative discussion can be derived from a mathematical approach. At any rate, Synergetics provides us with an insight into the possibility of steering complex systems using self-organization: namely by an appropriate choice of control parameters. Thus the conventional *direct* steering of each individual agent/part is replaced by some global measures ("indirect steering"). The concept of "nudging" of modern economic theories comes close to this concept (or is even identical with it).

3.3.2 The Paradigm of Pattern Formation

The starting point here is the Bénard (1900) instability. In this experiment, a thin layer of a liquid (e.g. oil) in a circular pan is heated from below and cooled from above. When the temperature difference between the lower and upper surface is small, heat is transported microscopically and no macroscopic motion of the liquid occurs. If,

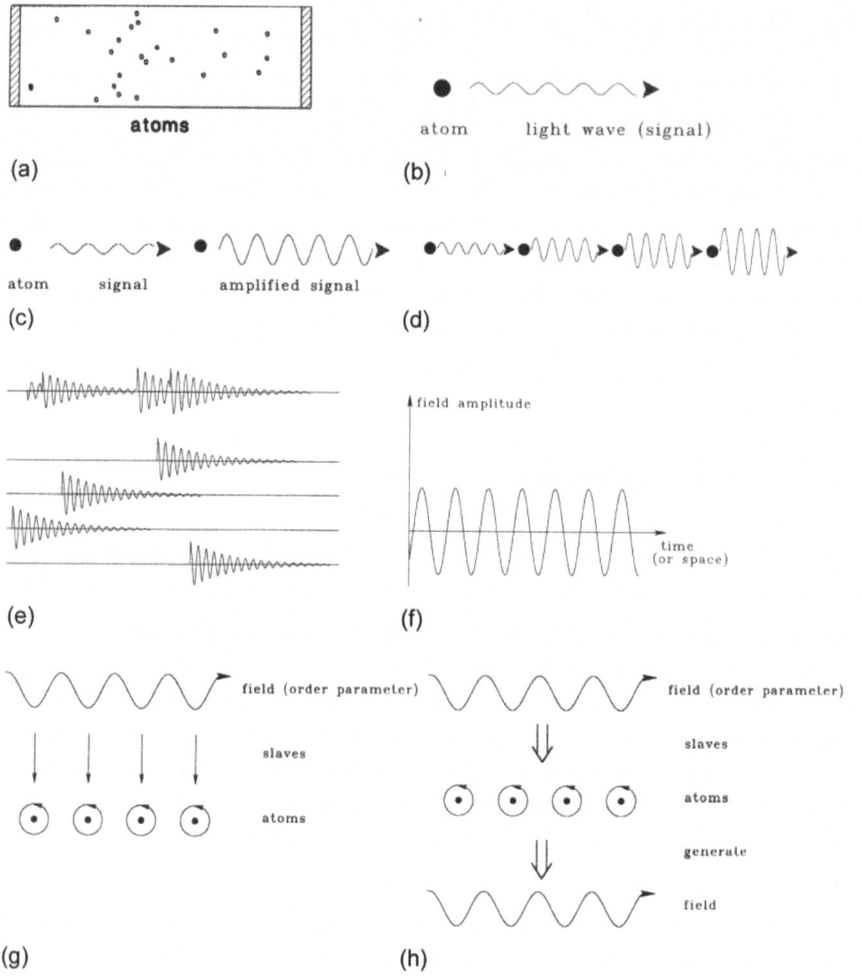

Fig. 3.2 The laser paradigm. **a** A typical setup of a gas laser. A glass tube is filled with gas atoms and two mirrors are mounted at its end faces. The gas atoms are excited by an electric discharge. Through one of the semi-reflecting mirrors, the laser light is emitted. **b** An excited atom emits light wave (signal). **c** When the light wave hits an excited atom it may cause the atom to amplify the original light wave. **d** A cascade of amplifying processes. **e** The incoherent superposition of amplified light waves produces still rather irregular light emission (as in a conventional lamp). When sufficiently many waves are amplified they strongly compete for further energetic supply. That wave that amplifies fastest wins the competition initiating laser action. **f** In the laser, the field amplitude is represented by a sinusoidal wave with practically stable amplitude and only small phase fluctuations. The result: a highly ordered, i.e. coherent, light wave is generated. **g** Illustration of the slaving principle. The field acts as an order parameter and prescribes the motion of the electrons in the atoms. The motion of the electrons is thus "enslaved" by the field. **h** Illustration of circular causality. On the one hand, the field acting as order parameter enslaves the atoms. On the other hand, the atoms by their stimulated emission generate the field, (Haken, 2006)

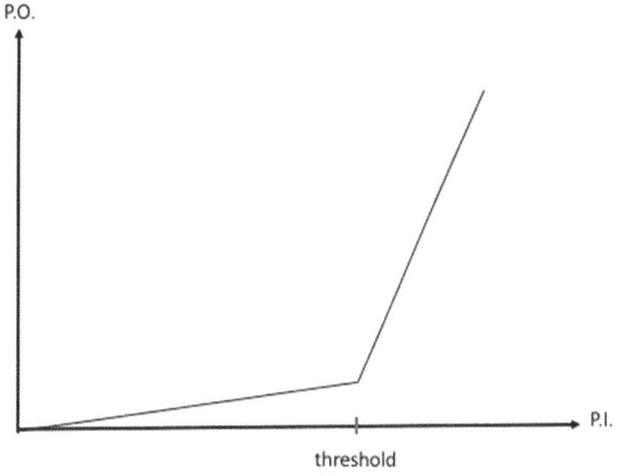

Fig. 3.3 The power output of the laser light wave (p.o.) versus power input (p.i.), e.g. power of injected electric current. The sharp increase of the slope proves that coherent action strongly enhances efficiency

however, the temperature difference exceeds a critical value, a macroscopic pattern emerges: a honey comb structure (Fig. 3.4 top left). When, in addition, also the border of the pan is heated uniformly, the structure changes qualitatively: the hexagons are replaced by a spiral (which can be one-or multi-armed) (Bodenschatz et al. 1992; Bestehorn et al. 1993). The temperature difference thus controls the macroscopic behavior of the system and functions as a control parameter.

As the control parameter is increased, the initial resting state becomes unstable and the liquid starts its motion. *Instability* thus shows up. Slightly above the instability

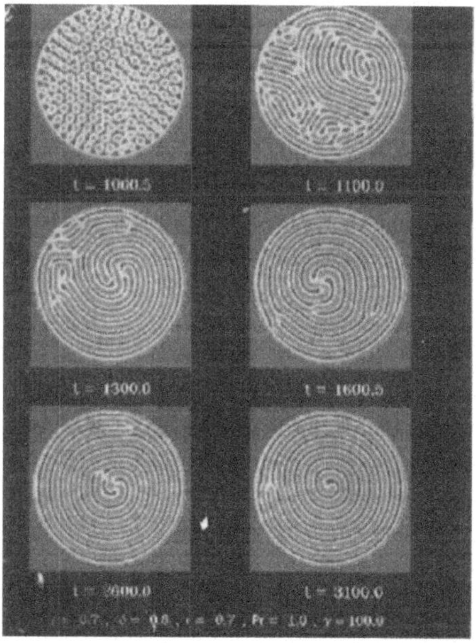

Fig. 3.4 Fluid layer heated from below: l.h.s.: formation of hexagons, r.h.s.: formation of spirals when the border is heated in addition (From Bestehorn et al. 1993)

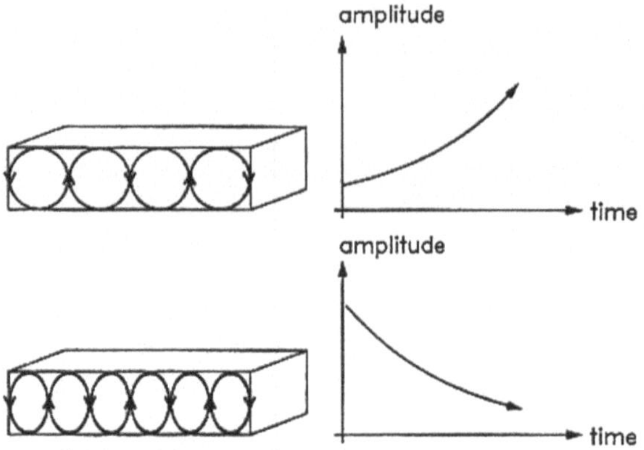

Fig. 3.5 *Left*: two different role configurations in a fluid. *Right*: The behavior of the amplitudes of these configurations in the course of time. While in one case the amplitude increases, in the other it decays

point, the system may undergo quite different collective motions of role configurations (Fig. 3.5). At the beginning the amplitudes of these role configurations are small and independent of each other. When they grow further, they start to influence each other—in some cases they compete until one configuration suppresses the others, in others, they co-exist and even stabilize each other. This happens in the formation of the honey comb structures. The amplitudes of the growing configurations act as order parameters that describe the macroscopic structure of the system (Haken 1996, p. 39).

The order parameters not only determine the macroscopic structure of the system, but also govern the space-time behavior of its parts. By winning the competition the order parameters enslave the many parts of the system to their specific space–time motion. This is a basic theorem of synergetics and as noted above, it has been termed *the slaving principle*.

In some cases, for example when the fluid is enclosed in a circular vessel, all directions for roll systems are then possible, each being governed by a specific order parameter. Which pattern will eventually be realized, depends on initial conditions. It is as if the system internally stores many patterns. This repertoire of patterns is not stored in a static fashion, but is dynamically generated anew each time. This property is termed *multistability*.

3.3.3 The Paradigm of Pattern Recognition

A typical experiment of pattern recognition can start as follows: a test person (or a computer) who has many patterns, of faces, city maps, etc., stored in memory, is offered a portion of one of the patterns. The task is to recognize the pattern—to decide what face, city map, etc. it is. According to synergetics, what happens is a process analogous to pattern formation as described above: at the start, the cognitive system of the person (or computer) is in a state of multistability as it enfolds many patterns, which coexist. When a few features or part of the pattern

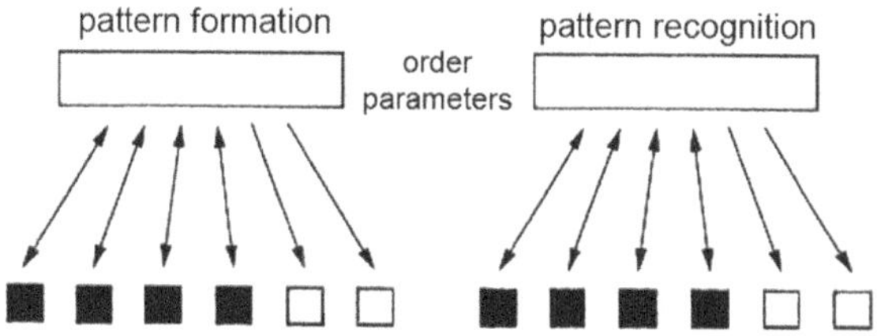

Fig. 3.6 Haken's (1979, 1991) analogy between pattern formation and pattern recognition. *Left:* a configuration of some parts of a system gives rise to an order parameter which enslaves the rest of the parts and brings the whole system to an ordered state. *Right:* a few features of a pattern shown to a person (or a computer) generate an order parameter which enslaves, and thus complements, the rest of the features, so that the whole pattern is recognized

is offered, several pattern configurations and their order parameters are formed by means of *associative memory*. The order parameters enter into competition and when a certain order parameter wins the competition and enslaves the cognitive system, the task of recognition is completed. This analogy, which is illustrated in Fig. 3.6, was first demonstrated by Haken in 1979 and has since become the basis for an intensive study of synergetics of cognition and of brain development and activity (Haken ibid. 1979, 1990, 1996). Figure 3.7 is a typical implementation with respect to face recognition, by means of the so-called *synergetic computer*.

As suggested by Portugali and Haken, the above conceptualization offers also an appropriate framework for the study of cognitive maps of cities, regions, and large-scale environments (Portugali 1990; Portugali and Haken 1992). The basic idea is that cities, regions etc. can be regarded as large-scale patterns, which can never be

Fig. 3.7 Pattern recognition of faces by means of the synergetic computer. *Top*: examples of prototype patterns stored in the computer, with family names encoded by letters. *Bottom*: when part of a face is used as an initial condition for the pattern recognition equations, their solution yields the complete face with the family name

seen in their entirety. As a consequence, the cognitive system constructs a whole cognitive map on the basis of only a partial set of features available to it. Thus it can be said that a partial set of environmental features offered to a person triggers a competition between several configurations of features and their emerging order parameters, until one (or a few) wins and enslaves the system so that a cognitive map is established.

3.3.4 Pattern Formation and Pattern Recognition in the City

The above analogy between pattern formation and pattern recognition provided the foundation for some of the major advances made in synergetics in connection with issues of cognition and brain functioning (Haken 1996). In light of these studies, Haken and Portugali (1995) suggested that the analogy between pattern formation and pattern recognition is specifically attractive to the study of cities. Cities can be perceived as complex, self-organizing systems, which are both physical and cognitive: individuals' cognitive maps determine their location and actions in the city, and thus the physical structure of the city, and the latter simultaneously affects individuals' cognitive maps of the city. In their preliminary mathematical model Haken and Portugali construct the city as a hilly probability distribution which is evolving, changing and moving as a consequence of the movement and actions of individuals (firms etc.). The latter give rise to the order parameters, which compete and enslave the individual parts of the system and thus determine the structure of the city. The significant and new feature of this exposition is that the order parameters enslave and thus determine, two patterns: one is the material pattern of the city, and the other is the cognitive pattern of the city, that is to say, its cognitive map(s). This is exemplified diagrammatically in Fig. 3.8.

One of the more interesting outcomes of the model is the set of *attention parameters*, which emerge by means of self-organization. In a state of multi-stability, or in case of an ambivalent pattern (e.g. 'vase or faces?', 'young or old lady?' in Fig. 3.9) they determine which aspect of the pattern is seen (i.e. attracts attention) first. This is of outmost importance in city dynamics. The city is full of patterns, yet individuals are attentive to only a few of them. The latter form the cognitive maps of the city and it is according to them that individuals and firms behave, take decisions and act in the city. In the model we investigate cases were one attention parameter dominates the dynamics and cases where no cross-attention is paid; that is, when two or more urban communities are cognitively not aware of each other. This situation entails the emergence and persistence of an urban cultural or socio-economic mosaic where a few coexisting attention parameters govern the dynamics.

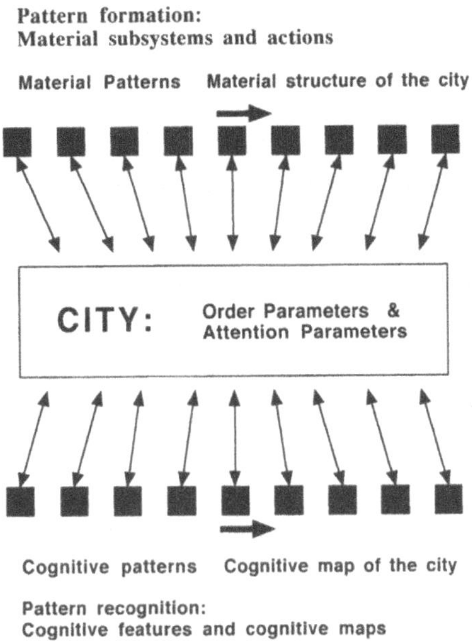

Fig. 3.8 The city as a self-organizing system that is at the same time both physical and cognitive. Its emerging order- and attention-parameters enslave the city's cognitive and material patterns, (Portugali, 2000)

Fig. 3.9 *Left*, vase of face? *Right*, young lady of old lady?

3.3.5 The Finger Movement Paradigm

The finger movement experiment was first conducted by Kelso (1984) and further modeled by Haken et al. (1985) to be known as *the HKB model*. The physiologist Kelso asked test persons to move their index fingers in parallel at the tempo of a metronome. At the beginning when the tempo of the metronome and the frequency of the finger movement were small this behavioral task could be performed quite well. Gradually the experimenter increased the speed of the metronome and the finger movement. Then suddenly, quite involuntarily, a switch to another kind of movement occurred, namely, to a symmetric movement (Fig. 3.10). The control parameter here was only the speed of the finger movement. This behavioral phase transition has been treated by means of synergetics in all details, including the so-called critical fluctuations and critical slowing-down.

As can be seen, this experiment illustrates self-organized behavior of an individual person. Another experiment conducted by Schmidt et al. (1990) indicates that the same happens in the case of two persons (Fig. 3.11). In the latter, two seated persons were asked to move their lower legs in an anti-parallel fashion and to watch each other closely while doing so. As the speed of the leg movement was increased, an involuntary transition to the in-phase motion suddenly occurred, in line with the Haken-Kelso-Bunz (ibid.) phase transition model (Haken 1996, pp. 87–90).

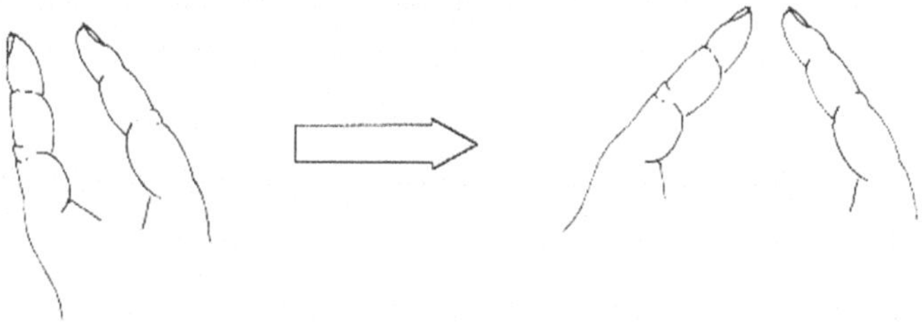

Fig. 3.10 Kelso's finger movement experiment. While initially people can move their index fingers in parallel, beyond a critical speed of the finger movements the relative position of the fingers switches involuntarily to the antiparallel, i.e., symmetric, position

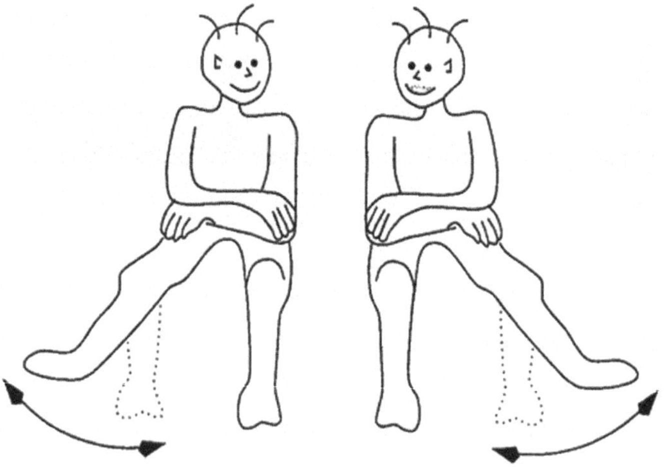

Fig. 3.11 The Schmidt et al. (1990) leg movement experiment with results identical to the above Kelso's experiment

This experiment is of special significance because it implies collective behavior—a phenomenon that plays an important role in urban dynamics—a case in point is pedestrians walking speed in cities of various sizes that we describe in Chap. 6.

3.4 The 1st and 2nd Foundations of Synergetics

As synergetics was applied to more and more areas, specifically to cognitive and brain processes, it was realized that the data needed for the bottom-up micro processes and for the top-down macro processes are always partial. As we've seen with respect to the bottom-up process of pattern recognition, it was possible to overcome this limitation by means of an analogy between the physical process of pattern formation (e.g. *Bénard convection*) to the cognitive process of pattern recognition. Such an approach that has been termed the *Synergetics' 1st Foundation,* is described in details in Chap. 5. As for the top-down approach, there was a need of making unbiased guesses on the state (or function) of the total system consistent with the known data. The approach to fulfill this need is termed the *Synergetics' 2nd Foundation* and is presented in Chap. 6.

References

Alexander, C., Ishikawa, S., & Silvestein, M. (1977). *A Pattern language.* New York: Oxford University Press.

Barabási, A.-L. (2016). *Network science.* Cambridge: Cambridge University Press.

Batty, M. (2013). *The new science of cities.* MIT Press Cambridge Mass.

Bénard, H. (1900). Les tourbillons cellulaires dans une nappe liquide. *Revue Générale des Sciences Pures et Appliquées, 11,* 1261–1271 and 1309–1328. See also: Les tourbillons cellulaires dans une nappe liquide transportant de la chaleur par convection en régime permanent. *Annales de Chimie et de Physique, 23,* 62–144 (1901).

Bestehorn, M., Fantz, M., Friedrich, R., & Haken, H. (1993). Hexagonal and spiral patterns of thermal convection. *Physics Letters A, 174,* 48–52.

Bodenschatz, E., Cannell, D. S., de Bruyn, J. R., Ecke, R., Hu, Y.-C., Lerman, K., & Ahlers, G. (1992). Experiments on three systems with non-variational aspects. *Physica D, 61,* 77–93.

Haken, H. (1979). Pattern formation and pattern recognition—An attempt at a synthesis. In H. Haken (Ed.), *Pattern formation by dynamical systems and pattern recognition* (pp. 2–13). Berlin/Heidelberg/New York: Springer.

Haken, H. (1990). *Synergetics of cognition.* Berlin/Heidelberg/New York: Springer.

Haken, H. (1991). *Synergetic computers and cognition.* Berlin, Heidelberg: Springer.

Haken, H. (1996). *Principles of brain functioning: A synergetic approach to brain activity, behavior and cognition.* Berlin/Heidelberg/New York: Springer.

Haken, H., & Portugali, J. (1995). A synergetic approach to the self-organization of cities. *Environment Planning B, 22,* 35–46.

Haken, H., Kelso, J. A. S., & Bunz, H. (1985). A theoretical model of phase transition in human hand movement. *Biological Cybernetics, 51,* 347–356.

Haken H. (2006) Information and Self-organization: A macroscopic approach to complex systems. (3rd enlarged ed.). Springer: Berlin/Heidelberg/New York

Hillier, B. (2016). The fourth sustainability, creativity: Statistical associations and credible mechanisms. In J. Portugali & E. Stolk (Eds.), *Complexity, cognition, urban planning and design.* Springer Proceedings in Complexity, pp. 75–92.

Karndacharuk, A., Wilson, D. J., & Dunn, R. C. M. (2014). A review of the evolution of shared (street) space concepts in urban environments. *Transport Reviews, 34*(2), 190–220.

Kelso, J. A. S. (1984). Phase transitions and critical behavior in human bimanual coordination. *American Journal of Physiology, 246,* R1000–R1004.

Portugali, J. (1990). Preliminary notes on social synergetics, cognitive maps and environmental recognition. In H. Haken & M. Stadler (Eds.), *Synergetics of cognition* (pp. 379–392). Berlin: Springer.

Portugali, J., & Haken, H. (1992). Synergetics and cognitive maps. In J. Portugali (Ed.), *Geography, Environment and Cognition*, pp. 111–130. A special issue, *Geoforum, 23*(2).

Portugali J. (2000). *Self-Organization and the City*. Springer.

Schmidt, R. C., Carello, C., & Turvey, M. T. (1990). Phase transitions and critical fluctuations in the visual coordination of rhythmic movements between people. *Journal Exposition Psychology: Human Perception Performing, 16*(2), 227–247.

Wittgenstein, L. (1953). *Philosophical investigations.* (Translated by Anscombe GEM) Oxford: Blackwell.

Zipf, G. K. (1949). *Human behavior and the principle of least effort.* Cambridge MA: Addison-Wesley.

Chapter 4
SIRN, IA and Their Conjunction (SIRNIA)

As noted in Chap. 2, cities are *dually hybrid complex*. Cities are dually complex, in that the city as a whole and each of its parts—the urban agents—is a complex system (Portugali 2011, 2016). Looked upon from the perspective of synergetics, we have two self-organization processes at two different scales: At the scale of the city as a whole we have a circularly causal process in which, the urban agents, by means of their interaction, give rise to the city in a bottom-up manner, while the city acting as order parameter enslaves (i.e., describes and prescribes) the behavior of its agents in a top-down manner. The other self-organization process takes place at the scale of each individual urban agent's mind/brain/body (MBB), constrained by the rules, laws, etc., of the city. Here the agent's MBB extracts information from its environment—the city, processes this incoming information with previously learned and memorized information, and on the basis of this interaction acts on the city. The SIRN process captures this process.

Cities are also *hybrid* complex systems in that they are composed of artifacts (buildings, roads, neighborhoods, cities, etc.) which are simple systems, and human agents, each of which, as just noted, is itself a complex systems. Urban agents thus interact not only among themselves, but also with single urban artifacts (buildings, roads,..) as well as with collective artifacts such as neighborhoods and whole cities. The process of IA employs information theory in order to capture the information exchange between the human agents and single or collective urban artifacts.

The notions of SIRN and IA were introduced in some details in the past (Haken and Portugali 1996, 2015; Portugali 2011, 2016); here our aim is to introduce them shortly and then to shed light on their conjunction, that is: SIRNIA. We thus first introduce each separately and then focus on the conjunctive SIRNIA process as it is associated with urban dynamics.

H. Haken and J. Portugali, *Synergetic Cities: Information, Steady State and Phase Transition*, Springer Series in Synergetics,
https://doi.org/10.1007/978-3-030-63457-5_4

4.1 A SIRN View on Urban Dynamics

4.1.1 SIRN

As the name indicates, the notion of SIRN is composed of Synergetics—Haken's (1983) theory of complex self-organizing systems, as introduced above in Chap. 3, and IRN—a process referring to cognitive tasks that cannot be executed by a single cognitive act and are thus implemented by a sequential interaction between internal representations constructed in the minds/brains of people and external representations constructed by them in the world in the form of utterances, texts, drawn figures, action, behavior and the like (Portugali 1996). A simple example of such a process is multiplication aided by a pencil and paper on which, one can externalize (i.e. write down) the numbers to remember (Portugali 2011, 153).

4.1.2 The Basic SIRN Model

The basic (or general) SIRN model (Fig. 4.1) symbolizing a complex self-organizing active agent that is subject to two flows of information: internal and external. The first is coming from the agent's mind/brain, in the form of ideas, fantasies, dreams, thoughts, imagination, emotions and the like, while the second comes from the 'world' via the senses, the agent's body and/or artifacts. The interaction between these flows gives rise to an OP that in line with the principles of synergetics, governs the agent's action and behavior, as well as the feedback information flow to the agent's mind. This 'feedback information flow' refers to the formation of internal

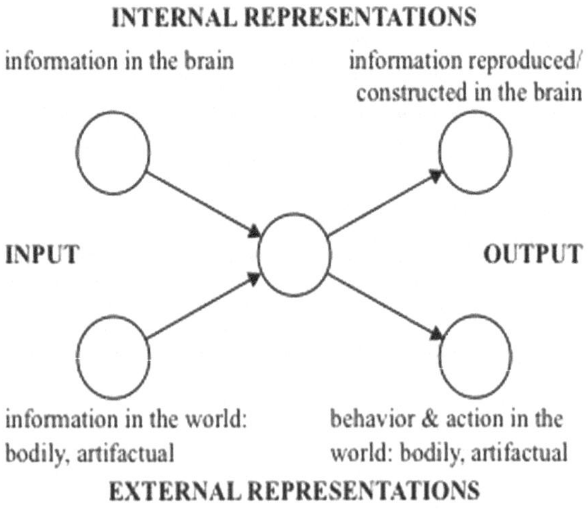

Fig. 4.1 The basic SIRN model as derived from the synergetic computer. SIRN symbolizes a self-organizing agent that on the one hand is subject to two forms of information (internal and external) and on the other actively constructs two forms of information, again internal and external

representations, such as images or learned patterns. The order parameters are determined by a competition along the lines of synergetic pattern recognition (Haken and Portugali 1996; cf. Chap. 3 above).

4.1.3 The Three SIRN Sub-Models

In order to apply our general SIRN model to specific case studies, we've derived from it three prototypical sub-models that refer to three principal cognitive contexts (Haken and Portugali 1996): the *intrapersonal*, the *interpersonal-sequential*, and the *interpersonal collective* (Fig. 4.2): The first refers to a solitary agent, the second to the sequential dynamics of several agents, and the third to the simultaneous interaction among many agents.

A typical intrapersonal process in the context of a city is design by means of sketching, as commonly practiced by architects and urban designers (Fig. 4.3). The designer starts with a given idea internally constructed and represented in the mind, draws it on a sketch paper (i.e., externalizes it), the drawn sketch gives rise to new ideas, problems or solutions, and so on in a play between internal and external representations.

A typical process of the interpersonal–sequential sub-model is the production of small-scale artifacts such as pottery objects (see Fig. 4.4): it starts with a certain design object (design 1) that already exists in the world. Designer 1 looks at design 1 and internalizes its form in his/her memory and then, based on this internalized representation produces design 2 as an external representation and so on. As can be seen, in this play between internal and external representations, designs 1 and 2 are mediated by designer 1's mind (body), designs 2 and 3 are mediated by designer 2's mind (body), and so on. Furthermore, since each pair in the sequence is mediated by a designer's mind (body), there is always the possibility of a copying mistake, that is, for a 'cultural mutation' and fluctuation (cf. Chap. 2). For a real-life example of this process at an urban scale consider the case of Tel Aviv balconies in Sect. 4.4.2, below.

The third SIRN model—the *interpersonal collective*—is also a model of urban dynamics (Haken and Portugali, ibid.; Portugali 2000, 2011). As illustrated in Fig. 4.2, each individual agent in the city is subject to internal input constructed by the agent's mind/brain, and external input which is the information afforded by the city as a "common reservoir". The interaction between these two forms of input gives rise to a competition between alternative decision rules that ends up when one or a few decision rules "wins", thus forming the OP that enslave(s) the system. The interaction between the actions of the many urban agents gives rise to the city's OP that enslaves and thus governs the actions of the individual agents.

As can be seen, the whole of the SIRN process is driven by the exchange of *information*: The agents extract the information conveyed by the city and then act on the city thus affecting its structure and consequently the information it further conveys

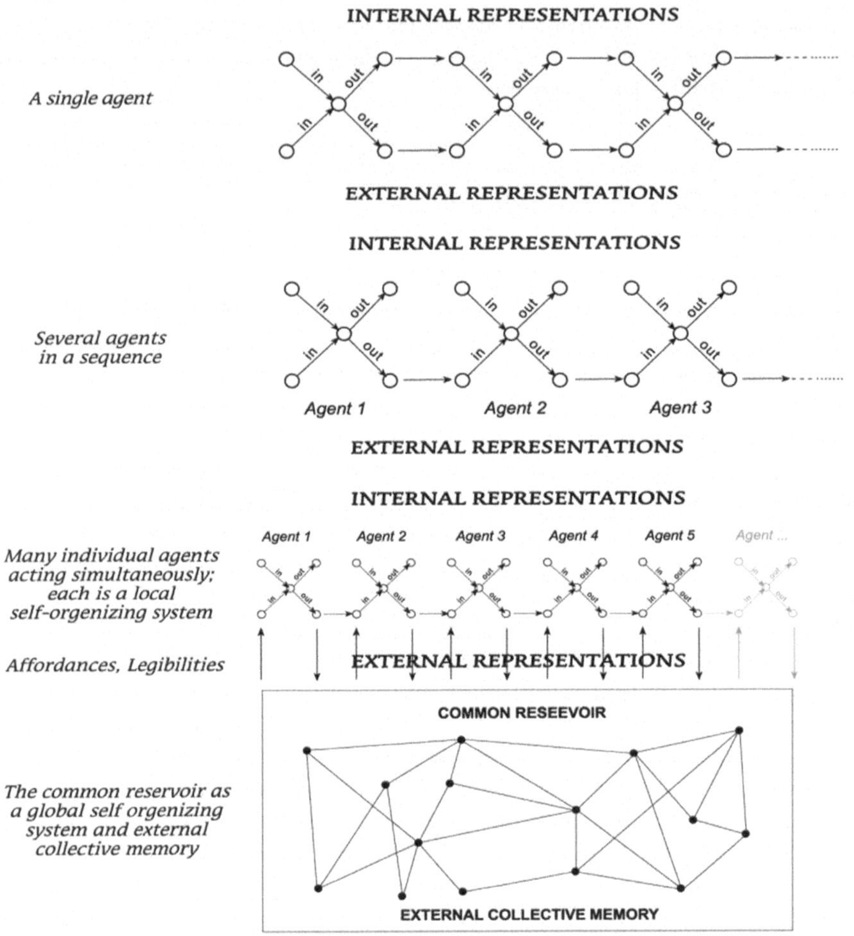

Fig. 4.2 The three SIRN sub-models: *top*, the intrapersonal. *Middle*, the interpersonal-sequential. *Bottom*, the interpersonal collective. Note that in this sub-model information and interaction between the urban agents are mediated by the common reservoir (e.g., a city)

and so on in circular causality. Furthermore, cities are not natural-organic complex systems; rather, as noted above, they are *hybrid complex systems* (Portugali 2016) some of whose elementary parts—the urban agents—are indeed natural-organic entities, while others, such as buildings, roads, neighborhoods, cities and metropolises are artifacts. In natural complex systems the parts interact and give rise to the global system; in the case of cities, the artificial parts cannot interact; their interaction is *mediated* by the natural parts—the human urban agents. As for urban agents, they partly interact with each other but as can be seen in Fig. 4.2, their interaction is mediated by the city as a whole. Such processes of interaction are based on the property that the artificial parts of the city at their various scales (buildings, neighborhoods

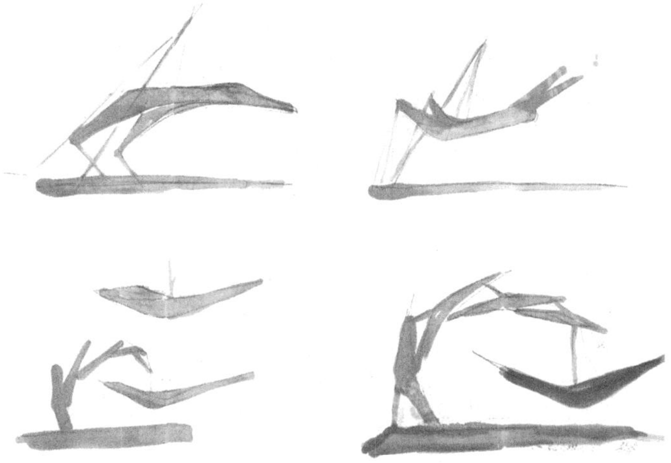

Fig. 4.3 A typical example is the sequence of design sketches by architect Santiago Calatrava

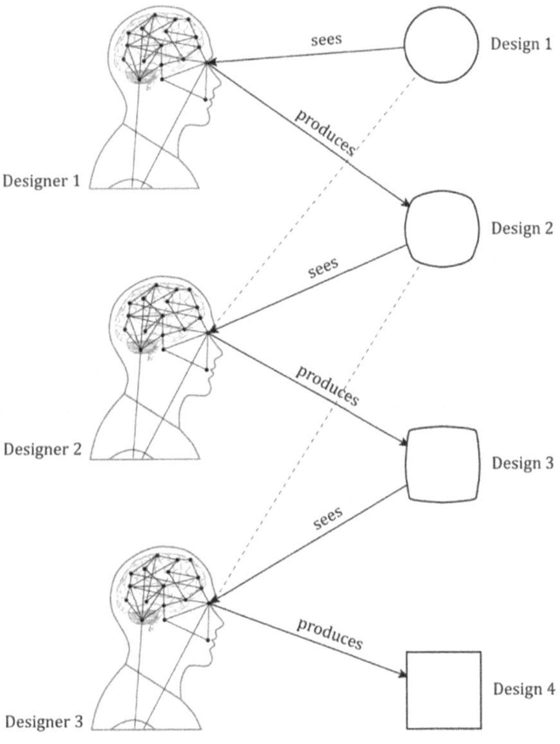

Fig. 4.4 The interpersonal–sequential sub-model. A case in point, the production of small-scale pottery objects

and whole cities) convey *information* that urban agents are capable of reading and on the basis of this information they then behave, interact, build, or act on, other artifacts. But then, *what is information?*

4.2 Information and the City

4.2.1 *Shannon Information*

Our immediate response to the question "*what is information?*" in the context of cities was to refer to Shannon's information theory (SHI) that since its first appearance (Shannon 1948; Shannon and Weaver 1949) has provided, and still provides, the foundation to any discussion of information. SHI deals with the capacity of *communication channels* to transmit *signals* of all kinds. It suggests that this capacity depends on the statistical properties of the signals, but not on their meaning. In this sense, channel capacity is a fixed physical quantity in each specific case, devoid of meaning. Shannon suggested several notions of information quantity of which the most common one is the *information bit* that defines information in terms of entropy:

$$i = -K \sum_k p_k \ln p_k, \tag{4.1}$$

where K is a constant related to the log_2. This definition allows us to calculate the (information) entropy $\boldsymbol{i}$ of any signal with a known $\boldsymbol{p}$, that is, the relative frequency (or probability) of distribution of symbols, distinguished by the index k.

4.2.2 *The Face of the City is Its Information*

One early application of SHI was in Gestalt theory, where it was used to show that "*good gestalt* is a figure with some high degree of internal redundancy" (Attneave 1959, 186). From the latter follows that different geometrical forms convey different quantities of information that can be quantified by means of SHI bits (e.g. Fig. 4.5).

Inspired by the above, in our "The face of the city is its information" (Haken and Portugali 2003; Portugali 2011, Chap. 8), we show that different elements/artifacts in the city, as well as different configurations of these urban elements, afford the perceiving urban agents different levels of information that can be quantified by SHI bits. For example, as shown in Fig. 4.6, when all buildings in a city are similar to each other (top line), information *i is low;* when they are different (second line down), *i is high* but hard to memorize; when landmarks (i.e. high rises) are added to a city with otherwise identical buildings, at different points (third line down), *i increases*; and when the high rises are grouped (bottom line), *i decreases.*

Fig. 4.5 Rotating the circle on the left four times by 90° conveys zero information bits, as the circle remains the same whatever its rotation. A circle in this respect is "a good gestalt." On the other hand, rotating an L-shape form four times by 90° conveys two information bits ($i = \log_2 4 = 2$), as 90° rotations could give rise to four different forms based on Algom 1986, p 115

Fig. 4.6 When all buildings are similar (top line), *i is low;* when they are different (second line down), *i is high* but hard to memorize; when landmarks are added, but separated from each other (third line down), *i is high*; and when they are grouped (bottom line), *i is low* Haken and Portugali, 2003

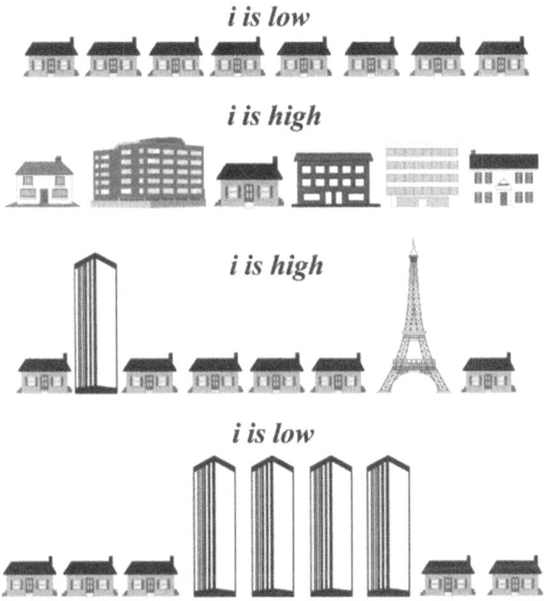

4.2.3 *Semantic Information Enters in Disguise*

One interesting outcome from our "face of the city" study was that despite the fact that SHI is defined independently of meaning, information with meaning, enters in disguise into its mathematical definition—a view originally suggested by Haken (2006) in his *Information and Self-Organization*. This is so since the choice of the index $\boldsymbol{k}$ in Eq. (4.1) above requires the categorization of urban elements such as buildings into, say, building styles (modern, postmodern and so on), functions (such as residential, offices, or industrial) or a combination of the two. And categorization, by definition, implies giving *meaning* to different urban elements or, in other words, *applying semantic information* (SI).

4.2.4 *Information Adaptation*

SHI and SI, are thus interrelated. The next step is to explore the role they play in the SIRN process, that is, in the interaction between the two flows of information—the one that comes from the city and the one that comes from the agent's mind/brain. In a recently published monograph (Haken and Portugali 2015) we show that these two forms of information are interrelated as aspects of a process of *information adaptation* (IA) in which, SI controls SHI, while SHI generates SI. Our canonical example here is the process of vision. Based on Hubel and Wiesel's seminal studies (1959, 1962, 1965) and on more recent findings (e.g. Livingstone 2002; Freiwald and Tsao 2010) and interpretations (Kandel 2012; Poggio and Serre 2013) we interpret the process as in Fig. 4.7: This process starts in the eyes by means of discrete, locally arranged receptor cells which send their signals (encoded by electrical pulses) as "data" to specific neurons that form a first layer of the brain. These neurons play a role analogous to that of the pixels of an image. But in contrast to inactive pixels they process the data further in a bottom-up manner, into local SHI of lines, corners and similar elements; this local SHI triggers a top-down process of synthesis ("reconstruction," in Kandel's words) that gives rise to global SI—to meaningful experience such as seeing and recognition. In the synthetic process, global SI, such as categories, interacts with quantitative Shannonian local information. In this interaction, SHI is adapted to SI by information inflation or deflation.

An example of information adaptation implemented by means of *information deflation* is the 'Kaniza triangle' illusion (Fig. 4.8 *left*), in which we see lines where there are none; our brain adds virtual line where no lines exist to mark out intersecting triangles. The "Olympic rings" illusion (Fig. 4.8 *right*) serves as an example of

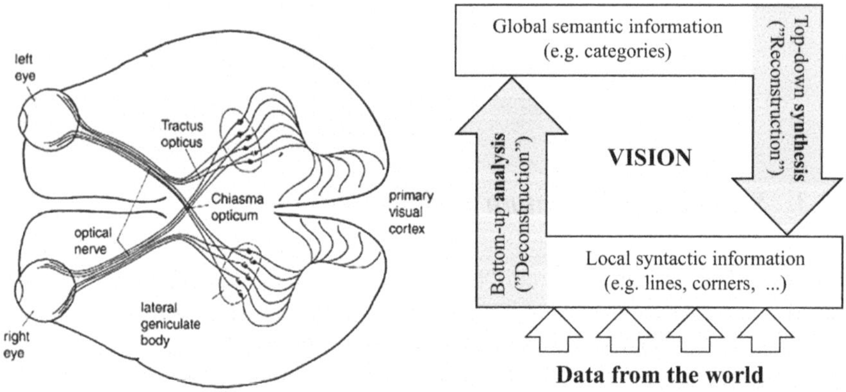

Fig. 4.7 The process of vision: *Left*, the links between the eyes and the brain; *right*, a schematic description. The process evolves as follows: data from the world is first analyzed by the brain, in a bottom-up manner; this local information triggers a top-down process of synthesis that gives rise to global information—that is, to seeing and recognition (see Haken and Portugali 2015)

Fig. 4.8 *Left*: the 'Kaniza triangle' illusion. *Right*: the "Olympic rings" illusion

information adaptation implemented by means of *information inflation*: we see five circles in superposition and overlook the many geometric forms of which this figure can also be composed.

4.2.5 Historical Note[1]

On first sight, the above dependence of SHI on categorization and thus on qualitative SI, was rather surprising. However, our further scrutiny into the origins of information theory revealed that information with meaning was part and parcel of Shannon's information/communication theory from its day one (Haken and Portugali 2015).

One of the main sources to most studies on information theory is Shannon and Weaver's (1949) book *The Mathematical Theory of Communication* which was composed of Shannon's paper from 1948 and of an extended version of Weaver's (1949) paper. In their introduction to their book Shannon and Weaver (1949, p. 4) recommend that Weaver's part should "... be read first" as it gives a " panoramic view of the field" and " includes ideas [...] for broader application". These ideas refer to the possibility of incorporating semantic and pragmatic information within the overall framework of Shannon's theory of communication. "Relative to the broad subject of communication", writes Weaver (Shannon and Weaver 1949, p. 4), " there seem to be problems at three levels. Thus it seems reasonable to ask, serially:

Level A. How accurately can the symbols of communication be transmitted? (The technical problem.)

Level B. How precisely do the transmitted symbols convey the desired meaning? (The semantic problem.)

Level C. How effectively does the received meaning affect conduct in the desired way? (The effectiveness problem.)".

[1]This section is taken from Haken and Portugali (2015), Sect. 1.2.

Weaver then suggests that " the theory of Level A is, at least to a significant degree, also a theory of levels B and C" and " that one's final conclusion may be that the separation into the three levels is really artificial and undesirable" (Shannon and Weaver 1949, p. 25). It is the generality of Shannon' s theory at Level A, writes Weaver, that makes it applicable to all kind of symbols (" spoken words, or symphonic music, or pictures") as well as to Levels B and C. Note that the above relations between Levels A, B and C reminds one of the relations between syntax, semantics and pragmatics in linguistics and computer science. The fact that Shannon and Weaver decided to co-author (and not to edit) their book indicates that Weaver accepts Shannon' s theory and that Shannon approves of Weaver' s interpretation and suggestions.

Shannon and Weaver thus laid the foundations for incorporating semantic and pragmatic information within the overall framework of Shannon's theory of communication. The process of IA by means of SI presented above was an attempt to follow their footsteps and show how SHI and SI are interrelated in the process of IA. In what follows we complete this task and shed light first on the process of IA by means of *pragmatic information* (PI) and then on the overall process of IA as an interplay between SHI, SI and PI.

4.2.6 Information Adaptation by Means of Pragmatic Information

So far we've considered one form of information adaptation, namely, adaptation by *semantic information* (SI). In our more recent study (Haken and Portugali 2016) we show that the city and urban elements in it convey another form of information with meaning—PI, that is, the action possibilities conveyed by a city or by its urban elements. Here we suggest that the process IA can thus take two forms: IA by means of SI, as above, and IA by means of PI as we show here.

In the above noted study we further suggested, that the notion of PI is very close to Gibson's (1979) notion of *affordance*, referring to the action possibilities afforded to animals, including humans, by the environment. "It implies", writes Gibson (ibid., 127) " the complementarity of the animal and the environment." Applied to a city scale we can say that a city conveys more/less PI in the form of jobs or housing opportunities, while at a smaller urban scale it can be said that an urban pavement conveys (affords) the PI of the action 'walking', while a road affords the action 'driving', whereas the steps to the Sacré-Cœur in Paris convey the PI action of 'climbing up' but also of 'seating' and enjoying the view of Paris (Fig. 4.9).

The above case brings forth a question regarding the relations between SI and PI: are the steps to the Sacré-Cœur 'steps' or 'benches'? Our answer: SI refers to the meaning per-se as commonly determined by the dominant society and its culture and language, whereas PI to the action possibilities afforded by an entity. Thus, in the case of the Sacré-Cœur we have *steps* (SI) that due to their size and form also afford *seating* (PI), even though they are not benches or chairs. Note that in the case

Fig. 4.9 The steps to the Sacré-Cœur in Paris convey also the PI action seating

Fig. 4.10 A chair that doesn't afford seating and a bicycle that doesn't afford cycling. From Jacques Carelman's *catalog of fantastic/unfindable objects*

of a chair, the outcome might be both SI and PI: depending on the circumstances, this object can be pattern recognized as a chair (i.e. SI) and at the same time as a 'seatable object' that affords the act of seating (i.e. PI). Can there be a non-seatable chair? Yes!: a miniature chair is still a chair despite the fact that it is not seatable. In Fig. 4.10 there are some examples of "non-affordable" objects taken from Jacques Carelman's *Catalog of fantastic/ Unfindable Objects.*

4.3 A Conjunction Between SIRN and IA

The general SIRN model (Fig. 4.1) symbolizing, as noted a complex self-organizing active agent that is subject to internal flow of information coming from the mind/brain and external information flow that comes from the 'world'. The interaction between these flows gives rise to an OP that governs, on the one hand, the feedback info to the mind/brain, while on the other, the agent's action and behavior. The interaction between the two information flows takes place in the agents MBB; but how? The answer is: by means of IA. The process of IA thus complements the SIRN process with the implication that the notions of SIRN and IA are in fact two aspect of a single process—SIRNIA.

4.3.1 SIRNIA

In the context of urban dynamics, the notion of SIRNIA thus refers to the interaction between urban agents and their city. It suggests that an urban agent is ongoingly subject to two flows (Fig. 4.11): A flow of *data* that comes from the city and a flow of *information* that originates in, and comes from, the agent's mind/brain/memory. By means of the latter, and in a bottom-up manner, the agent's brain first transforms the incoming data flow into a quantitative, "syntactic", Shanonnian information (SHI); this syntactic SHI triggers a top-down process that transforms the quantitative SHI into a qualitative semantic or pragmatic forms of information (SI and PI respectively). SI refers to the meaning per-se (i.e. 'this is a chair'), whereas PI to the action possibilities afforded by an entity, in a way similar to Gibson's (1979) notion of *affordances* (i.e. 'this object affords seating'). Figure 4.11 illustrates the SIRNIA process graphically.

For example, when looking at a ground floor of a city building, officially defined (by the planning law) as residential, one urban agent might see an apartment—i.e. the PI of living, while a second urban agent sees also the potential to transform this

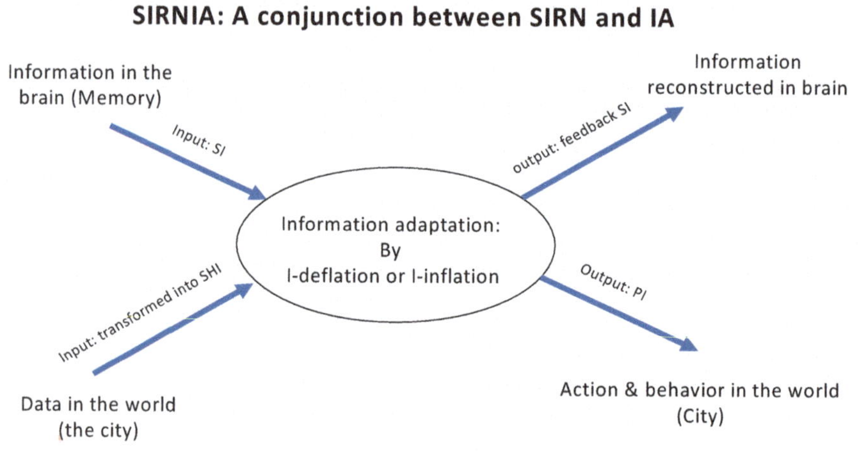

Fig. 4.11 The basic SIRNIA model. For details see text

apartment into an office, studio and/or a shop. Our first agent's action thus conforms with the city's plan/order parameter, while that of the second agent does not—it violates the city's order parameter /planning law. Our basic thesis is that such actions imply fluctuations or "urban mutations": small local deviations from the rule (the 'normal") that may lead to (i.e. trigger) macroscopic transformations. When these fluctuations occur during the long steady state periods, their effect is negligible and the "classical" scaling relations prevail; when in the unstable period, (that often leads to phase transition) we observe variations in scaling estimation (cf. Sects. 4.4 and 4.5).

4.4 Case Studies of SIRN-IA and the City

4.4.1 Lofts

As noted by Weaver in his collaborative book with Shannon (Shannon and Weaver 1949), SHI can be (intuitively) interpreted also as a measure of "freedom of choice". Applied to urban reality it has been suggested (Portugali 2016) that every urban element can be said to convey *objective* or *syntactic* SHI referring to its legal use. For example, for many years a large number of buildings in NYC were legally defined as a "Warehouse": and since according to the planning law of NYC they were to be used in one way only, the SHI of each of them was 0 bits:

$$I = \log_2 1 = 0\,\text{bits}$$

In the above (Portugali ibid.) it has further been suggested that every urban element conveys *subjective* PI referring to the specific way each urban agent perceives that urban element. For example, back in the 1960s, for an unknown NY artist desperately looking for a place to live and to work, motivated by a combination of imagination and pressing needs, the Warehouse probably conveyed different meanings (different PI) namely, a potential space for an apartment, a studio and even a shop. For this person the semantically (or pragmatically) determined SHI now became about 1.5 bits:

$$I = \log_2 3 \approx 1.5\,\text{bits}$$

This was probably the story of lofts in New York, London and other big cities around the world. As described by Kwartler (1998), in New York City, ad-hoc conversion of lofts in SoHo by individuals began in the 1960s, illegally and in contravention of both the New York City Zoning Resolution and Multiple Dwelling Law. Subsequently, this ad hoc activity was legitimized by revisions to both sets of regulations in 1982.

4.4.2 *Balconies*

A similar scenario unfolded in Tel Aviv (and subsequently across Israel) in what has been described as the "butterfly effect of Tel Aviv balconies" (Portugali 2011; Portugali and Stolk 2014). As illustrated and discussed in more details in Chap. 10 below, here, back in the late 1950s or early 1960s, an anonymous urban agent perceived the future state of his/her open balcony as a half room, planned a set of activities for it, designed the specific form of this half room and implemented the plan and design. Other people were attracted by this solution, closed their balconies too, and before long a process of innovation diffusion started that in a short period of time has transformed the whole of the urban landscape ("face of the city") of Israeli towns and cities. As in the case of lofts, it took several decades before the planning authorities legitimized closed balconies; in fact, in Israel (and by extension Tel Aviv) this happened very recently—during the year 2010. See Chap. 10 (Sect. 10.2.1) for a detailed description.

4.4.3 *Extreme Events*

During January–February 1991 Gulf War, Israel and specifically the Tel Aviv metropolitan area, were under Iraqi missile attacks. Thirty-nine Scud missiles were launched from western Iraq against Israeli targets during the five-and-a-half weeks of the war. Only six landed in populated areas, causing considerable property damage and the loss of a single life. Very soon the citizens learned that the missile attacks take place at night. As a response a new daily routine emerged—first by few, but then by many: people would leave the city in the evening to friends and relatives in the periphery of the metropolitan area and would return/commute back to the city in the morning to their daily work places and so on day after. This new pattern was strongly condemned by the Mayor of Tel Aviv, but in vain—people didn't listen to him and continued with their new daily routine until the end of the war. In retrospect many experts of extreme events consider this event as a proof for the title of Surowiecki's (2004) book that "*The wisdom of crowds: why the many are smarter than the few and how collective wisdom shapes business, economies, societies, and nations*".

4.4.4 *Streets' Corners*

Consider an urban block in a residential neighborhood in which each building is legally defined by the city's land-use plan and law as "residential". As a consequence, similarly to the case of lofts, the SHI conveyed by each building is 0 bits. However, each building also conveys PI referring to its potential use as determined, for instance, by its relative location in the block. From this perspective, corner buildings have

Fig. 4.12 See text

more exposure to by-passers and are thus potentially attractive also to commercial activities. Their potential PI determined SHI is thus 1 (compared to 0 of the mid-block buildings). Such a situation might be tempting to an urban agent looking for a non-expensive location for a shop or office: Namely, to rent/buy the flat and transform it into a shop or an office; as did, in fact, the urban agent who first transformed the ground floor flat at Zeitlin/Dubnov streets' corner in Tel Aviv into a shop, and as did a second urban agent who transformed the adjacent ground floor flat to an office (Fig. 4.12 *left*). In some cases, during time, the street at the blocks' shoulder will gradually be transformed into a commercial street, as happened in Bograshov st., Tel Aviv (Fig. 4.12 *right*).

4.4.5 Migration

The above three case studies were a consequence of a tension between an urban agent's residential or commercial needs and the PI locational possibilities conveyed by the city. Another typical solution to such a tension between the inhabitants' demand for space and the city's PI limit, is a migration move to a neighboring town. It is typical in such cases that the move is for residential purposes while the agent continues to work in the "big city" and uses the richness of its services (e.g. entertainment). Such a move that implies a link between the small suburban town and the city typically leads to the emergence of hierarchical urban system. Here too, the result is a structural change in the urban system.

4.4.6 IA at the Level of the City as a Whole

The above two examples refer to an IA at the level of a single urban agent. As shown in previous studies (Haken and Portugali 2015) the IA process is often implemented by acting on the city as a whole. To see how, consider Fig. 4.13.

Here, if all buildings in the imaginary city of Fig. 4.13 *left*, are similar to each other, the (SI determined) SHI conveyed by such a city is low; if all buildings in Fig. 4.13 *left* are different, the SHI such a city will convey is high. In the first case categorization into Uptown, Midtown, Downtown (Fig. 4.13 *right*) implies information inflation; in the second, information deflation. One of the outcomes of urban dynamics is an IA process (Fig. 4.14) that determines the appropriate balance between too much/little SHI conveyed by a city or neighborhoods in it. Such an IA process at the level of a city is usually implemented spontaneously by means of self-organization while in some cases by means of urban planning and design.

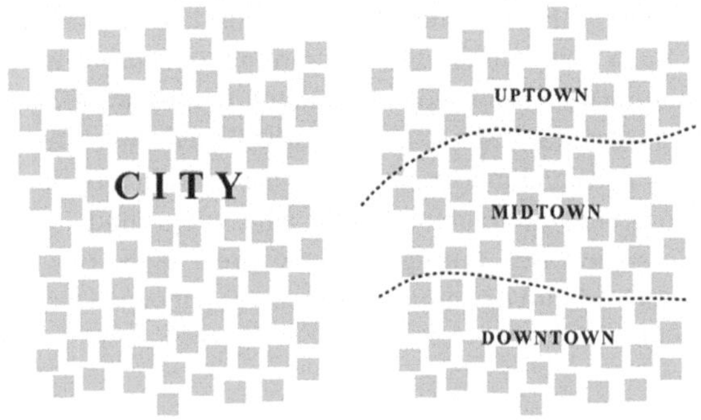

Fig. 4.13 The effects of categorizations a city into Uptown, Midtown and Downtown (Portugali, 2011)

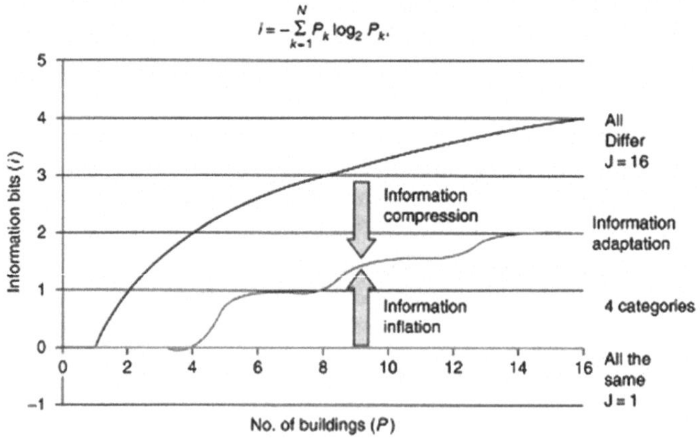

Fig. 4.14 The balance between too much/little SHI conveyed by a city. The equation on top refers to SHI as in Sect. 4.2.1 above. (Portugali, 2011)

4.5 Information Production: A SIRNIA View on Urban Dynamics

4.5.1 Information Production

From the perspective of SIRNIA, urban dynamics is a kind of production process—producing artifacts of all kinds, ranging from buildings, parks and roads to many other kinds of commodities, but also socio-cultural entities/products such as socio-economic or cultural areas (neighborhoods) and so on. These artifacts convey data from which urban agents extract SI and PI with their entailed SHI; then, on the bases of, or in response to, the extracted information, urban agents behave and act in the city. We term this SIRN-IA process the City's *information production* (IP). This IP process can be seen as, or gives rise to, the city's order parameter that affords a certain urban routine in all aspects of city life: ranging from commuting routinized patterns, to a steady production of residential, commercial, cultural and other goods and services.

Is it possible to quantify the amount of information produced by a city dominated as it is by a given order parameter? One possible positive answer is in line with the logic of Sects. 4.2.2 and 4.2.3 above, which are taken from our "face of the city" study (Haken and Portugali 2003): A city producing a single form of buildings (say, one story buildings only) conveys little PI and thus low/no SHI, that is, little choice for its inhabitants; a city producing several forms of buildings (say, one-, five- and fifteen-story high buildings), conveys much more PI (choice) and thus higher SHI. As we'll further note below, the same applies to the socio-economic and ethnic structure of a city.

The elementary principle in "the face of the city" is *urban diversity*: Low diversity refers to situations where the city's PI determined SHI is low, whereas high diversity to situations where the city's PI determined SHI is high. Applied to economic production it implies that a city producing one economic activity only conveys little PI (i.e. choice of economic actions) and its PI determined SHI is low; whereas a city producing several/many economic activities conveys high PI determined SHI (i.e. choice of economic actions). In the first case, the city is *vulnerable* to internal and external fluctuations; in the second, it is *resilient* to fluctuations. Example for the first case: Detroit's urban economy that was based mainly on the car industry (low diversity and thus low SHI) underwent a major urban decay (from a population of close to 2 million in 1950 to about 700.000 in 2013).

This tension between economic diversification (high SHI) versus economic specialization is a basic problem in decision-making of a company (or even of the financial market) and in our case of the economy of cities. The solution or answer to this tension is not quite simple, because diversification may increase e.g. fabrication costs etc., while as the case of "Detroit" shows, too strong a specialization may be fatal. As in the above case of the "face of the city …", in the economic reality of cities, this IA process that determines the balance between specialization versus diversification is rarely planned; more often than not, it is a self-organized emergent output of the city's dynamics. As we show next, the resultant vulnerability or resilience of the urban structure plays a central role in the evolutionary dynamics of cities (and as shown in Chap. 11, by implication their scaling laws and properties).

The same applies to the socio-economic and ethnic structure of a city: Low urban diversity in terms of socio-economic and/or ethnic diversity affords little choice, while high diversity much choice and so on. This is significant with respect to the current processes of immigration: Cities characterized by high socio-economic and cultural diversities are specifically attractive to new immigrants. These are usually the largest cities in a country.

4.5.2 Fluctuations

According to Synergetics, while in general random fluctuations of the individual parts of the system don't have an impact on the gross features of the system ("macroscopic" quantities) they play a major role in situations where the structure of the system changes qualitatively, i.e. its kind of order. This impact isn't taken care of by all deterministic approaches and requires, instead, a stochastic approach. Well-studied examples are in physics (theory and experiment), the nonequilibrium phase transition of lasers, Weidlich's (1971) model of public opinion and perhaps less known, the role of mutations in evolution. Here our aim is to study such deviations that as we show below, typify rapidly growing cities, which we conceive as selforganizing, adaptive systems. We show that in *rapidly* growing cities, fluctuations of the various aspects of a city such as population size, building activities, etc. play a major role violating macroscopic deterministic laws.

Synergetics thus indicates that fluctuations have an impact on the macroscopic structure of the system, when they occur in a situation or time when the global system is in a critical stage—when the system operated close to instability ("on the edge of chaos"). In the case of cities, as long as our SIRN-IA derived fluctuations occur far from instability, the city is resilient against such fluctuations. However, when as a consequence of some internal or external events the city is close to instability and thus vulnerable, small fluctuations previously enslaved by the city's order parameter can now lead the city into a phase transition.

In Sect. 4.4 above, we've explored several examples of IA events in the city at the level of single urban agents. In all, the decided solution was not in line with the city's routine, while in some case (e.g. NY lofts and Tel Aviv's balconies) the IA derived action was even *counter* the city's planning law and urban order. We suggest that from the perspective of urban dynamics such IA events can be interpreted as "mutations" or *local fluctuations* in a city's life. We further suggest, that IA derived fluctuations in cities are affected by the property that human agents have memory with the implication that a fluctuation that was enslaved is not forgotten and might have an effect on the city at a later stage.

With respect to the events of the lofts and balconies, if such law-breaking fluctuations had happened at different periods, their cities' planning systems would probably "enslave" them. This, in fact, was the case in Tel Aviv where at the 1940s attempts to close balconies were prevented by the city's planning authorities. But Tel Aviv balconies in the early 1950s and NY lofts in the 1970s took place when as a consequence of faster population increase, on the one hand, and space shortage, on the other, their cities were close to instability and thus vulnerable; at such a situation, a loft or a closed balcony fluctuation/solution attracted many residents of the city, many followed and imitated it (despite the planning law) and the result was a self-organized bottom-up process that leads to a phase transition in the global structure and face of the city.

From the above follows that urban growth proceeds stepwise as a sequential evolution of several logistic curves that imply long periods of steady states interrupted by short events of phase transition. It is thus typical for a city as an evolving complex system to pass through a series of non-equilibrium phase transitions driven by fluctuations as above, which nowadays might result from the introduction of electric cars, drones, autonomous driving and the like.

Concluding Notes

Cities as noted in Chap. 2 and above, are hybrid complex systems composed of artifact which are simple systems and human agents each of which is a complex system. As a consequence, in order to understand the complexity of cities we have to look at their complex components—the urban agents; for this purpose we have consulted the cognitive science—the science that deals with agents' cognition and behavior. This consultation indeed entailed significant insight to our understanding of human behavior in cities, yet at the same time it also exposed a problem: the cognitive science refrains from the study of artifacts. Now, artifacts as just noted, are one of the two basic components of cities. To our view, in the case of cities and in general, artifacts are integral components of the process of cognition. SIRN, IA and their conjunction SIRNIA, were specifically developed to capture the role of artifacts in processes of cognition and thus the way humans interact with artifacts, ranging from small artifacts to large-scale artifacts such as cities. As could be seen above, SIRNIA with its three sub-models captures the three main aspects of agent-artifact interaction: intrapersonal, interpersonal and collective which is in fact a cognitive theory of urban dynamics. As could be further seen above and as we show in subsequent chapters, in their interaction, urban agents respond not directly to the various urban artifacts, but to the information they convey; and, in this interaction they often create fluctuations that play a central role in the evolutionary dynamics of cities.

References

Algom D (1986). Perception and Psychophysics. Tel Aviv: Ministry of Defense Press (Hebrew).

Attneave, F. (1959). *Applications of information theory to psychology*. New York: Holt, Rinehart and Winston.

Freiwald, W. A., & Tsao, D. Y. (2010). Functional compartmentalization and viewpoint generalization within the macaque face-processing system. *Science, 330*(6005), 845–851.

Gibson, J. J. (1979). *The ecological approach to visual perception.* Boston: Houghton-Mifflin.

Haken, H. (1983). *Advanced synergetics.* Berlin/Heidelberg/New York: Springer.

Haken, H., & Portugali J. (1996). Synergetics, inter-representation networks and cognitive maps. In J. Portugali (Ed.), *The construction of cognitive maps*, pp. 45–67, Dordrecht, Kluwer Academic Publishers.

Haken, H., & Portugali, J. (2003). The face of the city is its information. *Journal of Environment Psychology, 23,* 385–408.

Haken, H., & Portugali, J. (2015). *Information adaptation. The interplay between shannon and semantic information in cognition.* Springer: Berlin, Heidelberg

Haken, H., & Portugali, J. (2016). Information and Selforganization. A unifying approach and applications. *Entropy, 18*, 197. https://doi.org/10.3390/e18060197.

Haken H. (2006). Information and Self-Organization: A Macroscopic Approach to Complex Systems, (3rd enlarged ed.). Springer.

Hubel, D. H. & Wiesel, T. N. (1959). Receptive fields of single neurons in the cat's striate cortex, *Journal of Physiology, 148*, 574–591.

Hubel, D. H., & Wiesel, T. N. (1962). Receptive fields, binocular interaction and functional architecture in the cat's visual cortex. *Journal of Physiology, 160,* 106–154.

Hubel, D., & Wiesel, T. (1965). Receptive fields and functional architecture in two nonstriate visual areas (18 and 19) of the cat. *Journal of Neurophysiology, 28*(229–89), 1965.

Kandel, E. (2012). *The age of insight: The quest to understand the unconscious in art, mind, and brain, from vienna 1900 to the present.* NY: Random House.

Kwartler, M. (1998). Regulating the good you can't think of. *Urban Design International, 3*(1), 13–21.

Livingstone, M. S. (2002). *Vision and art: The biology of seeing.* New York: Harry N Abrams.

Poggio, T., & Serre, T. (2013). Models of visual cortex. *Scholarpedia, 8*(4), 3516.

Portugali, J. (1996). Inter-representation networks and cognitive maps. In J. Portugali (Ed.), *The construction of cognitive maps* (pp. 11–43). Dordrecht: Kluwer Academic.

Portugali, J. (2000). *Self-organization and the city.* Berlin/Heidelberg/New York: Springer.

Portugali, J. (2011). *Complexity, cognition and the city.* Berlin/Heidelberg/New York: Springer.

Portugali, J. (2016). Interview in Lisa Kremer: What's the Buzz about smart cities? Tel Aviv University.

Portugali, J., & Stolk, E. (2014). A SIRN view on design thinking—An urban design perspective. *Environment and Planning B: Planning and Design, 41,* 829–846.

Shannon, C. E. (1948). A mathematical theory of communication. *Bell System Technical Journal, 27*(379–423), 623–656.

Shannon, C. E., & Weaver, W. (1949). *The mathematical theory of communication.* Illinois: Univ of Illinois Press.

Surowiecki, J. (2004). *The wisdom of crowds.* NY: Anchor Books.

Weidlich, W. (1971). The statistical description of polarization phenomena in society. *British Journal of Mathematical and Statistical Psychology, 24,* 251–266.

Chapter 5
Formalism I. Bottom–Up Approach: From Parts to Order Parameters

In this chapter we want to show how the concepts of Synergetics, Order parameters (OPs), slaving principle, circular causality, and control parameters, can be substantiated by a mathematical approach. In a first step, we have to choose the level of our description (reminiscent of the "Scale problem" of a cartographer).

As it transpires from Chap. 3, the systems and their parts may be of quite a different nature: parts as complex as humans, or parts as simple as single atoms (with only one light–emitting electron). Furthermore, the "parts" may be specific features, or indicators etc. that characterize a system. For a mathematical approach, they must be *quantifiable*.

5.1 Evolution Equations for the Individual Parts

In the case of a city, these indicators may be number of citizens, total income, number of flats, and so on. At a finer scale we may consider the number of citizens with different professions, or their local distribution, etc. We distinguish the individual "parts" by an index $j = 1, \ldots, J$ and denote their quantity (or intensity etc.) by q_j. We lump their entity into a "state vector q"

$$q = (q_1, q_2, \ldots, q_J) \tag{5.1}$$

that represents the state of our considered system. In general, the components q_j change in the course of time because.

(a) they are subject to externally fixed conditions
(b) they may influence each other
(c) they are exposed to internal and external chance events.

As usual, we denote the rate of change of q_j by

H. Haken and J. Portugali, *Synergetic Cities: Information, Steady State and Phase Transition*, Springer Series in Synergetics,
https://doi.org/10.1007/978-3-030-63457-5_5

$$\frac{dq_j}{dt} \text{ or shorter, } \dot{q}_j \tag{5.2}$$

This rate is determined by the sum of causes a, b, c, so that formally

$$\dot{q}_j = a + b + c \tag{5.3}$$

The explicit form of the contributions a, b, c, depends, of course, on the problem. For the moment being, a few hints may suffice.

(a) The "externally fixed conditions" are quantified by a small set of control parameters $\Lambda_1, \Lambda_2, \ldots, \Lambda_K$. For instance, the production rate of cars is proportional to the fixed energy input ("power supply"). In the case of the laser (Chap. 3) the light emission of its atoms is maintained by a power supply.
In this case, in Eq. (5.3) that refers to car production, (e.g. $j = 1$)

$$a = \Lambda_1 \tag{5.4}$$

where Λ_1 is the power supply times efficiency. Without any other effects the number of cars grows linearly with time. But cars don't run forever, i.e. their number decays at a total rate

$$-Kq_1 \tag{5.5}$$

where we consider K as a further control parameter Λ_2 (fixed by the technology of the time).

(b) Here we consider the interaction between the components q_j of (5.1). For instance, the production of a product j requires the cooperation of two other products, k and l. In this case, in Eq. (5.3) the term b reads

$$b_j = const.\ \ q_k q_l \tag{+}$$

where e.g., $j = 1, k = 2, l = 3$.

(c) We model the effect of chance events by means of very short, but intense pushes like the kicks of soccer players. A single kick at a time t_0 is modeled by means of Dirac's δ - function where

$$\delta(t - t_0) = 0 \quad \text{for} \quad -\varepsilon < t - t_0 < \varepsilon, \varepsilon \to 0, \quad \text{and} \quad \int_{-\infty}^{\infty} \delta(t - t_0)dt = 1 \tag{5.7}$$

The series of chance events, called random fluctuations $F(t)$, is represented by a sum over δ—functions at randomly selected times and with randomly selected strengths S_r

$$F(t) = \sum_r S_r \delta(t - t_r) \tag{5.8}$$

For practical purposes suffices is to know that the statistical average over times t_r and strengths S_r, $\langle \ldots \rangle$, is assumed to give rise to

$$\langle F(t) \rangle = 0, \quad \langle F(t) F(t') \rangle = Q\delta(t - t') \tag{5.9a}$$

where Q is the fluctuation strength.

In order to use F in Eq. (5.3) we must equip it with the index j used in (5.3) and generalize (5.9a) to

$$\langle F_j(t) \rangle = 0, \quad \langle F_j(t) F_k(t') \rangle = Q_{jk}\delta(t - t') \tag{5.9b}$$

In a number of cases the strength of the fluctuations depends on the variables q_j. In this case, we have to replace $F_j(t)$ by $g_j(q)F_j(t)$ (5.9b). For more details cf. below.

Summing up our insights and generalizing them in an obvious manner, we may formulate Eq. (5.3) as our basic evolution equation

$$\dot{q}_j = N_j(\Lambda, q) + F_j(t) \tag{5.10}$$

Or, using the representation (5.1)

$$\dot{q} = N(\Lambda, q) + F(t) \tag{5.11}$$

5.2 Complexity Reduction

Equations (5.10) and (5.11) are very general formulations of the behaviour of complex systems. Can we nevertheless devise a methodology that will allow us to find OPs, to cast the slaving principle into a mathematical form and to substantiate the concept of circular causality ? These questions lie at the root of Synergetics. To deal with them, in the first step, we assume:

(1) The impact of F in (5.11) can be ignored
(2) For a specific set of control parameters Λ a stable solution $q_0(t)$ to (5.11) is known. In the context of our book we assume that q_0 is time–independent (for more general cases cf. Haken (2004), *Synergetics. Introduction and advanced topics*). When we change one or several control parameters, the stability of q_0 may get lost. To check this, we superimpose on q_0 a small deviation δq, i.e. we put

$$q = q_0 + \delta q \tag{5.12}$$

where

$$q_0 = \left(q_1^0, q_2^0, \ldots, q_J^0\right), \quad \delta q = (\delta q_1, \delta q_2, \ldots, \delta q_J) \tag{5.13}$$

We insert (5.12) into (5.11) and linearize the r.h.s. with respect to δq. This leads us to a set of linear equations for the components δq_j, [cf. (5.13)].

The theory of linear differential equations with constant coefficients provides us with a remarkable result. There are J solutions all of which can be written as

$$\delta q^k = \exp(\lambda_k t) q^k \tag{5.14}$$

(Here we ignore degeneracies, where q^k may contain powers of time t).

In (5.14) q^k is a time-independent state vector

$$q^k = \left(q_1^k, q_2^k, \ldots, q_J^k\right), \quad k = 1, 2, \ldots, J \tag{5.15}$$

that represents a specific configuration of the features, indicators, etc., considered above when we formulated Eq. (5.11).

If one or several real parts of the characteristic values λ_k become positive, the deviation δq increases exponentially, indicating an *instability* of the state q_0. Here we consider only one real characteristic value with

$$\lambda_1 > 0 \tag{5.16}$$

Numerous studies have shown that systems may be "immune" (i.e. resilient) against a change of some control parameters, while they are very sensitive to changes of others especially at critical values where the instability sets in. In many cases of practical interest, even in systems with many components (e.g. in physics in lasers and fluids) close to instability points only one or a very small group of λs becomes positive. In order not to overload our presentation, we confine our analysis to only one λ_1 as indicated in (5.16).

Quite clearly, in real systems a deviation cannot increase forever. In fact, the nonlinearities inherent in (5.11) will serve for some limitation of growth. To deal with this "mechanism", we represent the solution $q(t)$ to (5.11) as a superposition:

$$q(t) = q_0 + \xi_1(t) q^1 + \sum_{k=2}^{J} \xi_k(t) q^k \tag{5.17}$$

And insert it in (5.11), including $F(t)$.

Leaving aside all technicalities (e.g. the dependence of q_0 on Λ) eventually we arrive at equations for the "amplitudes" $\xi_k(t), \quad k = 1, 2, \ldots, J$,

$$\dot{\xi}_1 = \lambda_1 \xi_1 + \widetilde{N}_1(\xi) + \widetilde{F}_1(t) \tag{5.18}$$

$$\dot{\xi}_k = \lambda_k \xi_k + \widetilde{N}_k(\xi) + \widetilde{F}_k(t), k = 2, \ldots, J \tag{5.19}$$

where $\xi(t) = (\xi_1(t), \ldots, \xi_J(t))$ and all $\widetilde{N}$ s are nonlinear functions that don't contain any constants or terms linear in ξ. Each $\widetilde{F}$ is a linear combination of the Fs in (5.11). Our further procedure is inspired by a comparison between λ_1 and the other λ_k s. Namely, we assume that besides $\lambda_1 > 0$ and $\lambda_k < 0, k = 2, \ldots, J$,

$$|\lambda_k| \gg \lambda_1 \tag{5.20}$$

Because the λ s have the dimension of an inverse time, (5.20) implies a "time-scale separation". To elucidate what this means, we consider an example: $k = 2$ and

$$\widetilde{N}_2 = C\xi_1(t)^2 \tag{5.21}$$

Because of the smallness of λ_1 [cf. (5.20)] we may assume that ξ_1 changes much more slowly than ξ_2 so that ξ_2 is subject to a practically time-independent "driving force" (5.21).

The general solution to Eq. (5.19), $k = 2$ then reads

$$q_2(t) = ae^{-\lambda_2 t} + \lambda_2^{-1} C\xi_1(t)^2 + G_2(t) \tag{5.22}$$

where a is a constant and $G_2(t)$ results from $F_2(t)$. When we treat processes that last longer than $1/\lambda_2$, we may neglect the rapidly decaying first term on the r.h.s. of (5.22). Using (5.9) and little mathematics we may calculate the correlation function

$$\langle G_2(t) G_2(t') \rangle = \lambda_2^{-1} \exp(-\lambda_2 |t - t'|) \tag{5.23}$$

In many cases we may *ignore fluctuations*. The result

$$\xi_2(t) = \lambda_2^{-1} C\xi_1(t)^2 \tag{5.24}$$

is easily obtained by putting.

$\dot{\xi}_2 = 0$ in (5.19). This is an example of the "adiabatic approximation" where we put $\dot{\xi}_k = 0$, $k = 2$, …, J, which transforms (5.19) into algebraic equations where $\xi_1(t)$ simply plays the role of a parameter. This approach is a powerful method when dealing with complex systems. Nevertheless, it is only the lowest approximation of an *exact* general procedure that allows us to express all $\xi_k(t)$, $k \geq 2$, as functions f_k of $\xi_1(t)$ at the same time!

Thus, written as formula

$$\xi_k(t) = f_k(\xi_1(t)) \tag{5.25}$$

where f_k depends on time-integrals over the fluctuating forces $\widetilde{F}$.

Equation (5.25) is the mathematical form of the *slaving principle*, and $\xi_1(t)$ is the *order parameter*. Let us make contact with Chap. 3 where we stated that the order parameter enslaves the individual parts. In the frame of the present Chapter, the "parts" are represented by the components q_j of q(5.1). Using (5.17) we obtain

$$q_j(t) = q_j^0 + \xi_1(t)q_j^1 + \sum_{k=2}^{J} f_k(\xi_1(t))q_j^k + G_j(v) \tag{5.26}$$

where we have included $G_j(t)$ as exemplified by (5.22).

This shows that the "behaviour" of the "part" q_j is determined by the OP ξ_1. Finally, when we insert (5.25) in $\widetilde{N_1}$ in (5.18) we obtain a closed equation for $\xi_1(t)$ alone:

$$\dot{\xi}_1(t) = \lambda_1\xi_1(t) + \widehat{N_1}(\xi_1(t)) + \widetilde{F_1}(t) \tag{5.27}$$

While Eq. (5.26) shows that the parts q_j are enslaved by ξ_1, Eq. (5.27) that determines the behaviour of ξ_1, has been brought about by the cooperation of the individual parts described by (5.18, 5.19): circular causality! Summing up we may state that the behaviour of a multi-component system can be described by that of a single OP. Actually, it is not difficult to derive equations for several OPs.

We turn to the discussion of typical OP equations. In many cases of practical interest it suffices to use only the leading term of a Tailor series expansion of $\widehat{N_1}$ in (5.27). We thus obtain well known equations, where we drop the index 1.

(1) The Verhulst equation (with a special fluctuating force)

$$\dot{\xi}(t) = \lambda\xi(t) - a\xi(t)^2 + \xi^{\frac{1}{2}}F(t), \xi(t) \geq 0, \tag{5.28}$$

and

(2) the laser amplitude equation

$$\dot{\xi}(t) = \lambda\xi(t) - b\xi(t)^3 + F(t) \tag{5.29}$$

In both cases, we may interpret λ as control parameter Λ.

Equations (5.28, 5.29) are the simplest example for the interplay between a control parameter, a nonlinearity, and a fluctuating force. Without the latter, the time-dependent and time-independent solution can be easily obtained and are not very spectacular: starting from any initial state $\xi \neq 0$, the system runs into a stable, time-independent state, e.g. $\xi = \lambda/a$ (5.28a). Equations (5.28) and (5.29) are intimately related to each other. To see this, we multiply (5.29) on both sides by $\xi(t)$ and put $\xi(t)^2 = n(t)$. We obtain

$$\frac{1}{2}\dot{n} = \lambda n - bn^2 + n^{1/2}F(t) \tag{5.30}$$

5.3 Fluctuations and the Fokker–Planck Equation

While fluctuations play a minor role, at least in general, they become important close to instability points where $\lambda \geq 0$. In this case, linear stability analysis amounts to neglect the cubic term in (5.29). The solution reads

$$\xi(t) = \int_0^t e^{-\lambda(t-\tau)} F(\tau)d\tau \tag{5.31}$$

While $\langle \xi(t) \rangle = 0$ because of (5.9), we obtain, again by use of (5.9),

$$\langle \xi^2(t) \rangle = (2\lambda)^{-1} Q(1 - e^{-2\lambda t}) \tag{5.32}$$

Letting $\lambda \to 0$, (5.32) acquires the simple form

$$\langle \xi^2(t) \rangle = Qt \tag{5.32a}$$

meaning that the fluctuating ξ may become arbitrarily large.

This phenomenon is well known in physics where it is called "critical fluctuations". In the context of urbanism, it has been referred to by Portugali (cf. Chap. 4). Here we want to treat it theoretically. In fact, a proper treatment of this effect is mathematically far more demanding because of the nonlinear term in (5.28) or (5.29).

We denote the Eq. (5.10) as Langevin equations because they are of the same type as the original Langevin equation of Brownian motion

$$\dot{\xi} = -\lambda\xi + F(t) \tag{5.33}$$

With a random force $F(t)$ obeying (5.9).

This stochastic process gives rise to a probability distribution $P(\xi)$ referring either to a process that is repeated again and again, or—an equivalent view—to an ensemble (of particles). Fokker and Planck were able to derive a differential equation for $P(\xi)$ that deals both with temporal evolution of P, i.e., $P(\xi; t)$, and its steady state corresponding to thermal equilibrium in the case of Brownian motion. Just for pedagogical reasons we quote their original equation

$$\dot{P}(\xi, t) = -\frac{\partial}{\partial \xi}(-\lambda\xi P(\xi; t)) + \frac{Q}{2}\frac{\partial^2}{\partial \xi^2} P(\xi; t) \tag{5.34}$$

This partial differential equation can exactly be solved which is not so interesting in the context of our book. Far more interesting is the fact that (5.34) can be generalized to a Fokker–Planck equation corresponding to (5.10). First, we consider the special case of a single variable ξ where (5.10) reads

$$\dot{\xi} = N(\xi) + F(t) \tag{5.35}$$

An explicit example is provided by (5.29). The (generalized) Fokker–Planck equation is obtained by replacing the linear term $-\lambda\xi$ in (5.34) by the nonlinear function $N(\xi)$ (e.g. $\lambda\xi - b\xi^3$)

$$\dot{P}(\xi; t) = -\frac{\partial}{\partial \xi}(N(\xi)P(\xi; t)) + \frac{Q}{2}\frac{\partial^2}{\partial \xi^2}P(\xi; t) \tag{5.36}$$

A further generalization has to be performed if the fluctuating force depends on the variable ξ such as in (5.28).

Mainly for the amusement for our readers we mention that a derivation of the corresponding Fokker–Planck equation has turned out to become rather tricky. To explain the situation, we remind the readers that we can visualize the fluctuations as individual kicks each of which changes the size of the variable $\xi(t)$ at some time. But which value of $\xi(t)$ must we use when such a kick happens? The Japanese mathematician Îto suggested to use $\xi(t)$ at a time just before the kick happens. The corresponding Îto-formalism (that differs from conventional calculus) leads to the Fokker–Planck—Îto equation

$$\dot{P}(\xi; t) = -\frac{\partial}{\partial \xi}(N(\xi)P(\xi; t)) + \frac{1}{2}G(\xi)\frac{\partial^2}{\partial \xi^2}P(\xi; t) \tag{5.37}$$

Most importantly, $G(\xi)$ is related to the factor $g(\xi)$ in (5.9b) and Q in (5.9),

$$G(\xi) = g^2(\xi)Q \tag{5.38}$$

The Russian mathematician/physicist Stratonovich suggested a different approach, namely to evaluate the "prefactor" of the "kick" at time t by means of the average $\frac{1}{2}(\xi(t + \varepsilon) + \xi(t - \varepsilon))$, $\varepsilon \to 0$. This leads us to the Fokker–Planck–Stratonovich equation. In the case (5.28) it looks somewhat more complicated than (5.37),

$$\dot{P}(\xi; t) = -\frac{\partial}{\partial \xi}((\lambda\xi - a\xi^2)P(\xi; t) + Q\frac{\partial}{\partial \xi}\left(1/2\xi^{1/2}\frac{\partial}{\partial \xi}(\xi^{1/2}P(\xi; t))\right) \tag{5.39}$$

Since $P(\xi; t)$ is a probability distribution, it has to obey for all times t the normalization condition

$$\int_{-\infty}^{\infty} P(\xi; t)d\xi = 1 \tag{5.40}$$

if $-\infty < \xi < \infty$
or

$$\int_{0}^{\infty} P(\xi; t)d\xi = 1 \tag{5.41}$$

if $0 \leq \xi < \infty$.

But why do we switch from Langevin equation to Fokker–Planck equation? Because the latter allow us to calculate the size of the fluctuation of ξ exactly by taking the nonlinear terms in (5.28, 5.29) into account!

Let us consider the individual steps. We start from the F.P. Equation (5.36). When we interpret $N(\xi)$ as a force, in analogy to physics, we may express it as the negative derivative of a "potential" (function) $V(\xi)$,

$$N(\xi) = -\frac{\partial V(\xi)}{\partial \xi} \tag{5.42}$$

In the special case (5.29), $V(\xi)$, reads

$$V(\xi) = -\frac{1}{2}\lambda\xi^2 + \frac{b}{4}\xi^4 \tag{5.43}$$

By use of (5.42) Eq. (5.36) acquires the form

$$\dot{P}(\xi; t) = \frac{\partial}{\partial \xi}\left(\frac{\partial V(\xi)}{\partial \xi} P(\xi; t)\right) + \frac{Q}{2}\frac{\partial^2}{\partial \xi^2} P(\xi; t) \tag{5.44}$$

Its exact, steady state solution

$$\dot{P} = 0 \tag{5.45}$$

reads

$$P(\xi) = N \exp\left(-\frac{2V(\xi)}{Q}\right) \tag{5.46}$$

The factor N is a normalization constant that can be calculated by means of the normalization condition (5.40) plots of $P(\xi)$ versus ξ are given in Fig. 5.1.

As (5.46) with (5.43) reveals, the size of the fluctuations of ξ remains finite because of the influence of the term

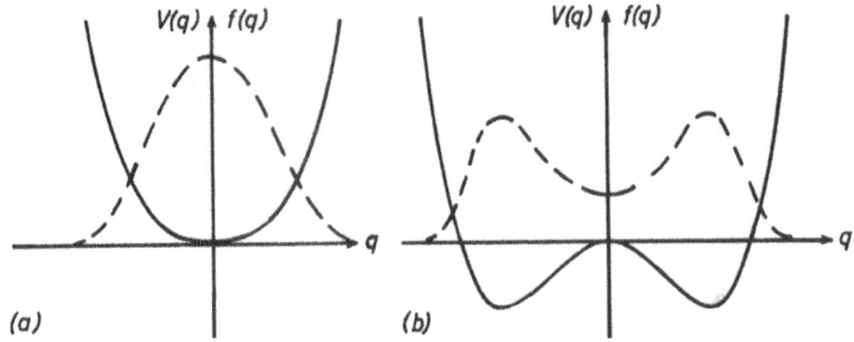

Fig. 5.1 Potential *V(q)* (solid line) and probability distribution $P(q) \equiv f(q)$ dashed line for (5.43) and (5.46), (**a**) $\lambda > 0$, (**b**) $\lambda < 0$

$$-\frac{1}{2}\frac{b}{Q}\xi^4 \tag{5.47}$$

which stems from the nonlinear term $-b\xi^3$ in the Langevin equation.

The size of the fluctuation of ξ is determined by the "width" of (5.46) around one of its maxima that are at

$$\frac{\partial V(\xi)}{\partial \xi} = 0, \quad \text{i.e.} \quad \xi = \mp\left(\frac{\lambda}{b}\right)^{1/2} \tag{5.48}$$

More precise statements can be made by the calculation of expectation values such as

$$\langle \xi^2 \rangle = \int_{-\infty}^{\infty} \xi^2 P(\xi) d\xi \tag{5.49}$$

We will provide the reader with explicit results in Sect. 10.4.

Because of the slaving principle, not only the order parameter undergoes fluctuations, but also the individual parts/components q_j are influenced. Their expectation values can be calculated by means of (5.26):

$$\left\langle \left(q_j - q_j^0\right)^n \right\rangle = \int \left(q_j - q_j^0\right)^n P(\xi_1) d\xi_1, \quad \text{n} = 1, 2, \ldots \tag{5.50}$$

The effect of "critical", i.e. enhanced fluctuations shows up also in the components.

5.4 The Fokker–Planck Equation for Many Variables

As we will demonstrate in the following Chap. 6 the Fokker–Planck equation for many variables may serve as a link between the bottom–up and top down approach. Here it may suffice to quote its form and a special steady state solution. The Fokker–Planck equation belonging to the "Langevin" Eq. (5.10) for the state vector q (5.1) reads

$$\dot{P}(q;t) = -\sum\nolimits_j \left(\frac{\partial}{\partial q_j} (N_j(q) P(q;t)) \right) + \frac{1}{2} \sum\nolimits_{jk} Q_{jk} \frac{\partial^2}{\partial q_j \partial q_k} P(q;t) \quad (5.51)$$

Here it is assumed that the fluctuating forces in (5.10) obey the statistical averages (5.9a). If the fluctuating forces depend on the variables q, and we apply the Îto-formalism, we have to replace the "diffusion" coefficients Q_{jk} by q-dependent functions $G_{jk}(q)$,

$$Q_{jk} \to G_{jk}(q) \quad (5.52)$$

In general, solving (5.51) becomes very difficult. Fortunately, however, there is an important special case where we can find the steady state solution to (5.51) explicitly. Namely, if we can derive the "forces" N_j from a "potential" $V(q)$,

$$N_j(q) = -\frac{\partial V(q)}{\partial q_j} \quad (5.53)$$

And

$$Q_{ij} = Q\delta_{ij} \quad (5.54)$$

The validity of (5.53) can be checked by

$$\frac{\partial N_j}{\partial q_k} = \frac{\partial N_k}{\partial q_j} \quad (5.55)$$

The conditions (5.54) requires that the fluctuating forces are statistically independent (stemming from different noise sources) and are of the same constant strength Q. Under these conditions, the steady state solution to (5.51) reads

$$P(q) = N \exp\left(-\frac{2V(q)}{Q} \right) \quad (5.56)$$

where N is the normalization constant, defined in analogy to (5.40), and (5.41), but over all $q_1, q_2, \ldots, q_J$.

Concluding Notes

In Chap. 3 we have presented and described the basic concepts of synergetics and its four paradigmatic case studies that together form the 1st Foundation of Synergetics. In this Chap. 5 we have presented the mathematical formalisms that complement and support the 1st Foundation—in general and in relations to cities. In particular the chapter introduces the Synergetic's formalism to the evolution of a complex system such as a city out of the interaction between its parts; the way OPs are formed and entail the processes of slaving and circular causality, and, the role of fluctuations when they occur close to instability points.

The mathematical formalism of Chap. 5 is of direct relevance to the study of agent-based models, that may form the basis for city planning ("scenarios") as well as policy. Since any more advanced models need nonlinear differential equations, computer solutions are required. Quite often it becomes difficult to interpret the large amount of data. In this chapter we have shown how to get insight into the mechanisms that lead from uncoordinated to coordinated actions. According to the conventional view, external orders are needed to establish coordinated actions. However, as we have shown in Chap. 3 by means of the swimming pool example, certain circumstances can enforce a highly coordinated behavior without any external organizer. Chap. 5 has put our intuitive approach on a solid mathematical basis. It may also serve as a guideline to solve and interpret agent based models. Our results shed light on a more recently discussed approach in economics known as "nudging"—an indirect steering of the behavior of agents (Thaler and Sunstein 2008; Sunstein 2019). Finally, as we shall see below, Chap. 5 prepares and paves the way to the subsequent chapters.

References

Haken, H. (2004). *Synergetics. Introduction and advanced topics*. Berlin, Heidelberg: Springer.

Sunstein, C. R. (2019). *How change happens*. Mit Press.

Thaler, R., & Sunstein, C. (2008). *Nudge: Improving decisions about health, wealth, and happiness*. New Haven: Yale University Press.

Chapter 6
Formalism II. Top–Down Approach: From Sparse or Big Data to Laws

Introduction: Synergetics Second Foundation
While in Chap. 5 ("Bottom–up approach: …") we presented an outline of the basic ideas of *Synergetics first foundation*, in the present Chapter. ("Top–down approach: ….") we deal with its second foundation. The bottom–up versus top–down tension accompanies complexity theories from the start, when the general tendency is clearly toward the bottom–up. Complex systems, it is commonly declared, emerge from the bottom–up. The exception is Haken's theory of Synergetics: It started with the LASER paradigm as a bottom–up process that gives rise to an OP; but then it adds that the OP enslaves the parts in a top–down manner, and so on in circular causality. However, as Synergetics was applied to more and more domains outside physics, specifically to the domains of cognition and brain functioning, a new genuine top–down view on complexity emerged—*Synergetics Second Foundation*.

Philosophically, the starting point of the Synergetics 2nd foundation is, that in terms of knowledge and data, complex systems are *inexhaustible*. To use Alfred Korzybski words: "whatever you say it is, it is not." A case in point that was studied by Synergetics is cognition and brain processes. Here, there exist top–down macro data as well as bottom–up micro data, when both are always partial. Regarding the micro, by means of analogy to LASER and to pattern formation (e.g. *Bénard convection*), a bottom–up formalism was devised to pattern recognition, for instance. This can be seen as Synergetics' major (e.g. 1st) foundation. As for the top–down approach, it is dictated, as noted, by the fact that since in the case of complex systems only a limited amount of data is known, there is need of making unbiased guesses on the state (or function) of the total system consistent with the known data. The appropriate mathematical tool to fulfill this need was Jaynes (1957) *maximum entropy principle* (MEP) and its extension to his maximum calibre principle. We will present the 2nd foundation below. However, before we do so we want to turn to the domain of CTC.

As in complexity theories at large, so in CTC too, the common view is that "cities emerge from the bottom–up". Here too, the exception is the *synergetic cities* approach that indeed agrees that cities emerge from the bottom–up, but then adds that once

H. Haken and J. Portugali, *Synergetic Cities: Information, Steady State and Phase Transition*, Springer Series in Synergetics,
https://doi.org/10.1007/978-3-030-63457-5_6

they come into being, cities enslave the behavior of their parts (the urban agents) in a top–down manner and so on in circular causality. To this we now want to add, that in terms of knowledge and data, cities as complex systems are *inexhaustible*, and, to rephrase Alfred Korzybski words: "whatever you say about cities, it is not." In what follows we thus suggest a preliminary discussion about *the 2nd foundation of synergetic cities*.

The previous discussion (Chap. 5) "From parts to order parameters", can be applied to the bottom–up approach of agent-based modeling that starts from "microscopic" evolution equations. The models make "ad hoc" *assumptions* on the behavior of, and relations between, agents. Nevertheless, as is shown in Chap. 5, general conclusions can be drawn on the emergence of collective behavior/structures based on the OP concept. In contrast, the present discussion (Chap. 6) starts from *measured data*. Depending on their kind and richness, we derive evolution equations either for OPs directly, or for an ensemble of agents. Then we may apply Chap. 5 to derive OPs and their equations.

What does it mean "to deal" with a complex system? In a first step we may collect data. But according to which criteria? What data? How many? How often? How precisely? In general, we may be confronted with an enormous amount of data ('big data"). How can we "digest"? Process them, draw conclusions? Do our data allow us to understand basic processes? In how far can we reduce complexity, establish simple(r) relationships?

In other cases we may have only sparse data at hand. In both cases big data/sparse data, can we develop some qualitative understanding or even a quantitative formulation? These are prerequisites for predicting future developments, planning and steering. How can we keep a system "alive" or robust against perturbations? How can we improve its efficiency, e.g. in cities minimizing energy consumption, waste production, pollution? In complex systems, quite often processes are intimately interwoven; e.g. relations between employment, energy consumption and "smartification".

In this Chap. 6 we want to develop a methodology that allows us to use the available data and to establish relations between them by means of a general formalism. By invoking results of Chap. 5 we show how we can reduce the complexity of a system. Our point of departure is Shannon information (SHI).

6.1 Statistics, Probability, Shannon Information. A Brief Reminder

Probability theory originates from a theory on gambling. Simple examples are tossing a coin (with two outcomes: head or number) or rolling dice (with 6 outcomes: number of eyes, 1, 2, 3, 4, 5, 6). Each throw is called a trial, leading to an event. In how far can we make predictions on future events based on past events? A central concept is relative frequency defined as:

relative frequency $= \frac{n_s}{n}$

where

n_s number of trials with successful outcome,
n total number of trials.

For a small number n, the relative frequency may vary considerably. It is assumed that for large n, the relative frequency approaches a fixed value. It is used to quantify predictions on the probability p of the outcome of a specific event

$$p = \lim_{n \to \infty} \frac{n_s}{n} \tag{6.1}$$

Since in practice we can't make infinitely many trials, we resort to sampling. A well-known example is polls to make predictions on the outcome of an election. This amounts to use an inter-(or extra)-polation. How many samples are needed for reliable predictions is dealt with by specific approaches we will not present here.

The possible outcomes are characterized by specific features which we distinguish by a quantifier q. In the case of a coin, we may put.

head: $q = 1$, number: $q = 2$.

The weight of persons, suitably discretized, may be such a q. Quite generally we assume that features can be quantified—even in psychology with scales by convention,

e.g. "sad": $q = 1$.

"happy": $q = 2$.

Or more detailed grades.

Another example is political opinion:

Pro/contra a specific project.

By use of the label q, we may write (6.1) as

$$p_q = \lim_{n \to \infty} \frac{nq}{n} \tag{6.2}$$

Since in our later applications we will treat q as a variable (that may even acquire continuous values) we will write $p(q)$ instead of p_q.

The probabilities $p(q)$ obey the normalization condition

$$\sum_{q=1}^{Q} p(q) = 1,\ Q\text{: total number} \tag{6.3}$$

For later use, we introduce the quantity

$$-\log_2(p(q)) = i(q) \tag{6.4}$$

which is called "information of the event q". Another name is: "surprisal", because $i(q)$ is large for small $p(q)$ (indicating a rare event).

In many practical cases we don't need the probabilities $p(q)$ explicitly but rather the mean and variance of q that are defined by Mean:

$$\overline{q} = \sum_{q=1}^{Q} p(q)q \tag{6.5}$$

Variance:

$$\overline{q}^2 - \overline{q}^2 = \sum_{q=1}^{Q} p(q)q^2 - \overline{q}^2 \tag{6.6}$$

Average value of a function of $q, f(q)$

$$\overline{f} = \sum_{q=1}^{Q} p(q)f(q) \tag{6.7}$$

Shannon information S is an average over the information (6.4)

$$S = -\sum_{q=1}^{Q} p(q)\log_2 p(q) \tag{6.8}$$

In our book we use the natural logarithm ln instead of $\log_2$

$$S = -\sum_{q=1}^{Q} p(q)\ln p(q) \tag{6.9}$$

Note that (6.9) and (6.8) differ by a numerical factor only that drops out in later basic results.

There may be cases where the outcome of a trial must be characterized by kinds of two different features, e.g. the height and weight of persons. These two kinds are differentiated by q_1, q_2. The corresponding *joint probability* that a specific set q_1, q_2 is found in a trial is denoted by $p(q_1, q_2)$. It obeys the normalization condition

$$\sum_{q_1=1}^{Q_1} \sum_{q_2=1}^{Q_2} p(q_1, q_2) = 1 \tag{6.10}$$

Clearly this concept can be generalized to any kind of M different features q_1, q_2 … q_M. We use the shorthand notation

$$q = (q_1, q_2, \ldots, q_M) \tag{6.11}$$

So that

$$p(q) = p(q_1, q_2, \ldots, q_M) \tag{6.12}$$

In generalization of (6.5)–(6.7) we may define correlations, e.g.

$$\overline{q_1 q_2} = \sum_{q}^{Q} p(q) q_1 q_2 \tag{6.13}$$

To finish our reminder of basic concepts of probability theory, we recall the definition of conditional probability $p(q_1|q_2)$. With $q = (q_1, q_2)$,

$$p(q_1, q_2) = p(q_1|q_2) p(q_2) \tag{6.14}$$

This means that we can always split the joint probability $p(q_1, q_2)$ into a product of the conditional probability $p(q_1|q_2)$ times the probability $p(q_2)$. Often, $p(q_1|q_2)$ is interpreted as the probability that "q_1" happens under the *condition* that "q_2" happens, or under the *hypothesis* that "q_2" happens.

Because $p(q_1, q_2)$ is the probability that q_I and q_2 happen simultaneously and thus q_1 and q_2 play a symmetric role (in this specific interpretation), we may equally well use the decomposition

$$p(q_1, q_2) = p(q_2|q_1) p(q_1) \tag{6.14a}$$

Note that the functional dependence of $p(q_1)$ on q_1 in (6.14a) may be quite different from that of $p(q_2)$ on q_2 in (6.14) as indicated by the different indices 1 or 2. The relations (6.14) and (6.14a) allow us to equate their right hand sides so that, after some light rearrangement, we arrive at the famous *Bayes' Theorem*

$$p(q_2|q_1) = p(q_1|q_2) \frac{p(q_2)}{p(q_1)} \tag{6.14b}$$

We will discuss its significance later. After the preparations we may turn to deal with the basic question: How to determine $p(q)$?

In a first step we have to select the appropriate features and their quantifiers 1, 2, ..., M. To elucidate some important implications we discuss city maps. A simple way is to treat such a map as an image and apply to its analysis a typical procedure used in image/pattern recognition (see below). Using a rectangular grid we decompose the image into its pixels. At present, we ignore the problem of choosing the appropriate scale. We label the pixels by an index l and attribute to each pixel a discretized *grey* value q_l, so that $p(q)$ depends on $q = (q_1, ..., q_L)$.

Quite clearly, when dealing with cities, the grey values are hardly of any value to us. Rather we must attach to each location a quantifier that tells us which kind of artifacts (buildings, streets, or parks, etc.) are at location "l". Should we distinguish between different kinds of uses, styles, …? And how detailed? It is here, where semantics come into the definition of $p(q)$ and thus into the value of Shannon information. After having fixed "q" how can we determine $p(q)$?

(1) By observation, counting, etc. but in all cases by sampling.
(2) By means of models dealing with processes underlying the change of $p(q)$ in the course of time.

Examples are: birth/death processes, migration of people within a city, or in/out of a city, change of opinion, or of profession, spread of epidemics, etc. One important theoretical tool is the master equation (Weidlich 2002). It reads

$$\frac{dp(q)}{dt} = \sum_k w(q;k)p(k) - p(q)\sum_k w(k;q) \tag{6.15}$$

The expression $w(q;k)$ is the transition probability per unit time, e.g. the number of transformations from state "k" to state "q" in one hour/ or day/ etc. In practice, (6.15) cannot be solved explicitly (with few exceptions) so that (6.15) is replaced by *mean value* equations [cf. (6.5)]. By contrast, in this chapter we will be concerned with methods that allow us to determine $p(q)$.

6.2 From Average to Probability Distributions. Jaynes' Maximum Information Entropy Principle

We start from Shannon information (6.9), i.e.

$$S = -\sum_q p(q)\ln p(q) \tag{6.16}$$

where

$$q = (q_1, q_2, \ldots, q_l) \tag{6.17}$$

and

$$\sum_q \text{ stands for } \sum_{q=1}^{Q_1} \cdots \sum_{q_l=1}^{Q_L} \tag{6.18}$$

Note the generalization of (6.9)–(6.16)!

Jaynes' principle consists in maximising S under given constraints

$$f^k = \sum_q p(q) f_q{}^k, k = 1, \ldots, K \tag{6.19}$$

In addition to the normalization condition

$$\sum_q p(q) = 1 \tag{6.20}$$

On the right hand side of (6.19), the $f_q{}^k$ are "suitably" chosen functions of q. Actually, the art of applying Jaynes' principle precisely consists in a "suitable" choice! In the case of a single quantifier, $q = 1,\ldots, Q$, the constraints might be.

$$k = 1 : f_q{}^1 = q, k = 2 : f_q{}^2 = q^2. \tag{6.21}$$

(Note that "2" of $f_q{}^2$ is an upper index!).

On the left hand side of (6.19), the f^k are experimentally measured quantities. In the case (6.21), f^1 is just the experimentally measured mean value of q. To determine the maximum of S under the constraints (6.19) and (6.20) we use the method of Lagrange multipliers $\Lambda_0, \Lambda_k, k = 1, \ldots, K$.

The central result reads

$$p(q) = \exp(\Lambda_0 + \sum_{k=1}^{K} \Lambda_k f_q{}^k) \tag{6.22}$$

In our above example

$$p(q) = \exp(\Lambda_0 + \Lambda_1 q + \Lambda_2 q^2) \tag{6.23}$$

The Lagrange parameters can be determined by inserting $p(q)$ in (6.19) and (6.20). We may write (6.22) in the form

$$p(q) = \exp(-V(q; \Lambda)) \tag{6.24}$$

where

$$-V(q; \Lambda) = \Lambda_0 + \sum_{k}^{K} \Lambda_k f_q{}^k \tag{6.25}$$

We may interpret $V(q; \Lambda)$ as representing a mountainous landscape where $V(q)$ in the height of a mountain (or valley) at the position q. Note that q need not be 2-dimensional, but may "span" a high-dimensional space. But nevertheless, this visualization serves our purpose. V provides us with a "bird's eye view". The minima of V, the position of its valleys are the most probable configuration of $q = (q_1, q_2, \ldots, q_L)$. While $V(q)$ determines the steady state (time—independent) probability distribution, we may attach to it a dynamics. To this end we interpret (6.24) as the steady state solution to a Fokker–Planck equation of the form (5.51) of Sect. 5.5 which belongs to a stochastic process whose fluctuating forces obey

$$<F(t)> = 0, <F(t)F(t')> = 2\delta(t-t')\delta_{lk}$$

with $Q = 2$.

6.3 Gaussians

The form of the probability distribution function (6.23) is strongly reminiscent of a Gaussian. In fact, it becomes a Gaussian if q can be treated as a continuous variable $-\infty < q < \infty$. In this case, we can replace

$$\sum_q \ldots \text{ by integral } \int_{-\infty}^{\infty} \ldots dq$$

The probability distribution $p(q)$ standing under the sum becomes $\widehat{p}(q)dq$, where $\widehat{p}(q)dq$ is a probability *density*. Keeping these different interpretations in mind, in the following we will write $p(q)$ instead of $\widehat{p}(q)$ and speak—somewhat sloppy—of "probability".

The constraints corresponding to (6.21) read

$$\langle q^k \rangle = \int_{-\infty}^{\infty} p(q)q^k dq = f^k, k = 1, 2 \tag{6.21a}$$

Note that "k" under the integral denotes an exponent "power", whereas at f^k it is a mere index. Since we seek the solution to (6.21a) and the normalization condition in the form (6.23), the constraints can be explicitly calculated which makes the use of Gaussians very appealing. Actually, whenever statistics or error estimations play a role in science (including psychology) and technology, the use of Gaussians is ubiquitous even in some Deep Learning machines. For later use we present the resulting equations for Λ_0, Λ_1, Λ_2.

Normalization

$$\frac{\pi}{\alpha}^{1/2} \exp\left(\Lambda_0 + \frac{1}{4}\frac{\Lambda_1^2}{|\Lambda_2|}\right) = 1, \alpha = |\Lambda_2|^{1/2} \tag{6.21b}$$

Constraints [by use of (6.21b)]

$$k = 1 \langle q \rangle = \frac{1}{2} \frac{\Lambda_1}{| \Lambda_2 |} = f^1 \tag{6.21c}$$

$$k = 2 \langle q^2 \rangle = \frac{1}{2 | \Lambda_2 |} + (\frac{1}{2} \frac{\Lambda_1}{| \Lambda_2 |})^2 = f^{(2)} \tag{6.21d}$$

Use of (6.21c), we may replace (6.21d) by

$$\langle q^2 \rangle - \langle q \rangle^2 = f^{(2)} - f^{(1)2} \tag{6.21e}$$

where the left hand side is the standard deviation, while $\langle q \rangle$ is the mean.

Thus the insight provided by Jaynes' principle is: whenever only mean and variance of a probability distribution are known, the best guess is a Gaussian!

Here we want to scrutinize in how far we can extend our approach to several (or many) $q_1, \ldots, q_L$. Our discussion will be somewhat "technical" so that a speedy reader may skip it.

The functions appearing in the constraints (6.19) read

$$f_q^k = q_k, \, f_q^{mn} = q_m q_n \tag{6.26}$$

So that (6.22) becomes

$$p(q) = \exp\left(\Lambda_0 + \sum_k \Lambda_k q_k + \sum_{mn} \Lambda_{mn} q_m q_n \right) \tag{6.27}$$

The sums run over all k, m n, from 1 till L.

The matrix (Λ_{mn}) is symmetric so that its characteristic values λ_μ, $\mu = 1, \ldots,$ are real. To evaluate the integrals over (6.27) and the constraints *explicitly* we must diagonalize (Λ_{lm}). Thus by use of transformed variables ξ_μ the bilinear term of (6.27) becomes

$$\sum_{\mu=1}^{k} \lambda_\mu \xi_\mu^2 \tag{6.28}$$

Because Jaynes' principle is based on *unbiased* guesses, we cannot exclude positive λ_μ s. But in such case the Gaussian integrals diverge! Thus we have to invoke constraints containing higher powers of q. Because some ξ_μ may be positive, we must include terms of at least forth order.

6.4 When in Rome Do as the Romans Do: Pedestrians' Behavior in Cities

Pedestrian movement is probably the most salient aspect of humans' behavior in cities. In 1976 Bornstein and Bornstein (1976) have published a paper showing correlation between population size of cities and walking speed of pedestrians in these cities (Fig. 6.1); this, as part of their attempt to study the impact of urbanization on the pace of life.

Subsequent studies have supported and elaborated these findings (Walmsley and Lewis 1989; Levine and Norenzayan 1999). More recently the issue appeared once again, this time, however, in the context of complexity theories of cities as part of an attempt to show that "many properties of cities from patent production and personal income to electrical cable length" as well as pedestrians walking speed, "are shown to be power law functions of population size with scaling exponents, β, that fall into distinct universality classes." (Bettencourt et al. 2007). In this section we suggest to interpret behavior in general and behavior in cities in particular as a form of information adaptation.

The pedestrian's behavior is a property that emerges out of the interplay between SHI and SI (Chap. 4). For example, when a newcomer settles in a city, s/he observes the other citizens and makes (hopefully unbiased) guesses on their behavior. In other words, s/he uses Shannon information, maximizes it under the observed constraints (e.g. average velocity etc.). This allows her/him to determine the attractors, i.e. the PI that instructs him/her on how to behave in accordance with the general behavior.

To arrive at an explicit mathematical model, we use the methods of Synergetics 2nd Foundation. The observable and measurable quantities are the pedestrians' mean

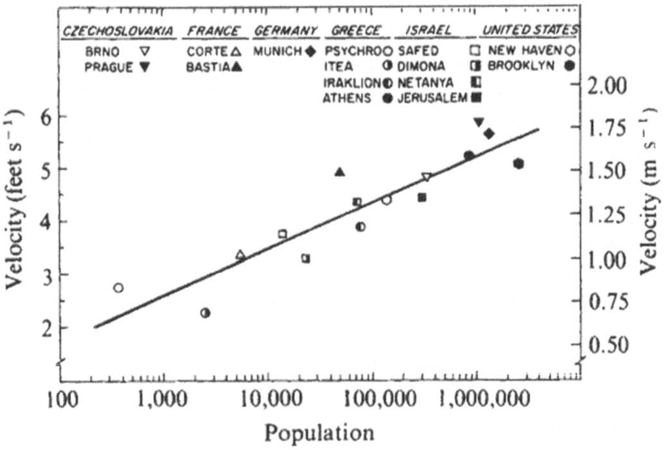

Fig. 6.1 The Bornsteins' correlation between population size of cities and walking speed of pedestrians in these cities

walking speed and its variance. The velocity ξ has all the properties of an order parameter:

1. it describes a property of the total system: here velocity distribution and its most probable velocity
2. it is brought about by the internal system dynamics
3. it enslaves the behavior of the individual parts: the velocity ξ determines, how many people $N(\xi)$ move at this velocity (on the average)
4. it is influenced by control parameter(s); here, city size
5. a change of control parameter induces a macroscopic change: here change of velocity.

To apply Jaynes' Maximum Information Entropy Principle, we need the constraints

$$\langle \xi \rangle, \langle \xi^2 \rangle$$

which are equivalent to mean and variance. Thus we obtain the probability distribution

$$p(\xi) = N \exp(-V(\xi))$$
$$V(\xi) = \lambda_1 \xi + \lambda_2 \xi^2$$

where the Lagrange parameters λ_1, λ_2 depend on city size as is suggested by the observations. Clearly, λ_1, λ_2 play the role of control parameters which must be related with city size. To get some insight, we seek that velocity ξ at which $p(\xi)$ acquires its maximum. This condition yields

$$\xi = \lambda_1/(2\lambda_2) \tag{6.28a}$$

Since $p(\xi)$ must drop to zero for $\xi \to \infty$ we require $\lambda_2 > 0$ and since $\xi > 0$, we require $\lambda_1 > 0$. But, importantly, as glance at Fig. 6.2 reveals, the ratio (6.28a) linearly increases with city size. It will be interesting to find data on the dependence of variance on city size so that we can determine both parameters λ_1, λ_2 independently.

The synchronization urge

Clearly, our new citizen doesn't use algorithms, but when we try to translate her/his *intuitive action* into the language of information processing, we may arrive at our IA interpretation. Or, put differently, when we were to devise a "citizen" robot, we would equip its brain with our IA algorithm.[1] (See below).

But what is the psychological-cognitive origin of this synchronization urge? Why would/should our newcomer synchronize behavior with the other inhabitants and why

[1]There is presently an interesting debate on the relations (or virtue) intuition/ algorithm (as e.g. applied to medical treatment in stroke units) by Gigerenzer (2015). He thinks that in important cases intuition (or heuristics) is better than algorithms.

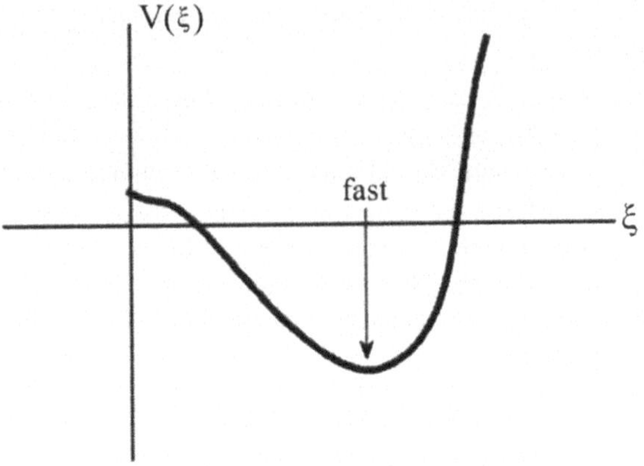

Fig. 6.2 Potential V versus walking speed

do they synchronize their walking speed in the first place? The answer comes from synchronization/coordination dynamics as in the finger movement paradigm and Schmidt's et al. legs movement experiments and HKB interpretation (cf. Chap. 3). As is well experienced and recorded, people walking together (who know each other) tend to synchronize pace etc. (and also their speed). Intuitively, probably the same might happen when many anonymous people in high density are walking to the same direction—they will give rise to an order parameter that will enslave their walking speed etc.

But 'why the walking speed in large cities is faster? There have been here several suggestions or rather speculations, ranging from the assumption that pedestrians try to avoid "social interference" (Borenstain 1976) and 'sensory overload' (Milgram 1970) that rises with the size of cities, to suggestions that people try to save time whose economic value is higher the larger the city is (Levine 1999).

To the latter we might add the following: behavioral movement in cities might be divided into productive (in workplace one produces and earns money) and non-productive (movement to workplace, i.e. commuting), which is often considered as a "waste of time". With few exceptions (siesta), the larger the city the longer is the non-productive time (journey to work). In small cities where everything is nearby, there is no waste of time, but in large cities it is a problem. In the latter, as part of their attempt to minimize the waste of time associated with the movement to work, people move faster.

6.5 Pattern Recognition

Why do we need pattern recognition in urbanism? We want to recognize structures and their changes, development, trends, etc. "Structure" is meant in a broad sense: land use, traffic, population, … This requires a cooperation between cognition, allometry, modeling etc. In this section we deal with pattern recognition from a general perspective. What is pattern recognition? How does it work? Pattern recognition is based on associative memory. A telephone book provides us with a simple example. It associates a specific telephone number to the name of a person. Quite generally on an abstract level, associative memory amounts to the completion of data. We may distinguish between *two phases*: (1) *Learning* and (2) *recognition.*

(1) *Learning.* There are a variety of approaches. We mention only two examples: (a) We establish a complete list (e.g. telephone book). (b) Sampling. The selection of samples depends on our objectives and realistic conditions. Let's consider face recognition as our objective. We want to recognize the faces of a preselected group of persons. "Realistic conditions" mean that we must recognize them in spite of different illuminations, even if they are partly hidden, at different distances (i.e. independent of their size) and so on. For each individual person a set of samples must be chosen that (hopefully!) includes typical situations. Eventually for each face one or two prototype patterns are constructed (front and side view).

(2) *Recognition.* Here a "test/ pattern" is shown which is a face under realistic conditions. The question is: Which prototype pattern does this test pattern belong to? In the following we present a brief outline of the "algorithm" that underlies the synergetic computer for pattern recognition, and that may serve as a model for pattern recognition by humans.

6.5.1 *Learning*

We superimpose a grid over an image. We distinguish the resulting pixels by an index l ("l": location) and attach a grey value q_l to each pixel. In the case of a city, we may interpret q_l also as a set of feature quantifiers as discussed above. Here and in the following all indices at the sums as l, m, n… run over all integers from 1 till L. We perform a slight preprocessing. We replace q_l by $q_l^{'} = q_l - \bar{q}$, where $\bar{q} = \frac{1}{L}\sum_l q_l$. Then we normalize the vector $q' = (q_1', q_2', \ldots, q_L')$ by means of the substitution $q' \to q'' = Nq'$ so that $(q'')^2 = 1$. We then drop the dashes". If needed we may make the patterns q invariant against displacements, rotation and scale in two dimensions. To apply Janyes' principle we use, in accordance with the preceding section, the constraints (besides normalization)

$$f_q^{nm} = q_m q_n \tag{6.29}$$

$$f^{mnrs} = q_m q_n q_r q_s \tag{6.30}$$

$$l_q = d\Big(\sum_m q_m^2\Big)^2 \tag{6.31}$$

where d is a scaling parameter, and e.g.

$$f^{mn} = \langle q_m q_n \rangle \tag{6.32}$$

Note that the averages $\langle \ldots \rangle$ must cover both the faces (objects) to be recognized but also their "realistic situation" (see above). When we insert (6.29)—(6.31) in (6.22) (with appropriately chosen indices k) and observe (6.25), we arrive at an explicit expression for $V(q;\Lambda)$ of the form

$$-V(q;\Lambda) = \Lambda_0 + \sum\nolimits_m \Lambda_{mn} q_m q_n + \sum\nolimits_m \Lambda_{mnrs} q_m q_n q_r q_s - \Lambda' \left(\sum\nolimits_m q_m^2\right)^2 \tag{6.33}$$

The Lagrange parameters Λ_0, Λ_{mn}, Λ_{mnrs}, Λ' are determined by constraints corresponding to (6.29)–(6.31) [note (6.32)] and the normalization condition

$$\int d^L q \exp(-V(q;\Lambda)) = 1 \tag{6.34}$$

The existence of (6.34) requires $\Lambda' > 0$ (6.33). As a detailed analysis reveals, in the generic case, the matrix elements Λ_{mn} are proportional to f^{mn}

$$\Lambda_{mn} = C f^{mn} \tag{6.35}$$

where C plays the role of a Lagrange parameter. Because f^{mn} is symmetric,

$$f^{mn} = f^{nm} \tag{6.36}$$

the matrix f^{mn} can be digonalized.

Thus the quadratic form

$$\sum_{mn} q_m f^{mn} q_n, \; m, n = 1, \ldots, L \tag{6.37}$$

can be transformed into

$$\sum_{\mu} \lambda_\mu \xi_\mu^2, \; \mu = 1, \ldots, L \tag{6.38}$$

where the λ_μs are characteristic values of the matrix (f^{mn}).

The transformation from (6.37) to (6.38) is achieved by means of putting

$$q_m = \sum_{\mu} \xi_\mu v_m^\mu = 1, \ldots, L \tag{6.39}$$

$$v^{\mu} = (v_1^{\mu}, \ldots, v_L^{\mu}) \tag{6.40}$$

The v^{μ}s (6.40) are the characteristic vectors of the matrix (f^{mn}).

They can be identified with the prototypes with label $\mu = 1, \ldots, L$. In the case of supervised learning, this label can be fixed by means of a marker (e.g. a person's name in face recognition or just μ *itself*) from the beginning, whereas in unsupervised learning its attachment may pose a problem. To elucidate it, we consider two cases.

(a) all characteristic values λ_{μ} are different. According to linear algebra, the characteristic vectors are uniquely determined and can be used as prototype patterns which we may arbitrarily distinguish by an index μ.
(b) Some or all λ_{μ} s coincide, say λ_1 and λ_2. Then any linear combination $av^1 + bv^2$ is also a characteristic vector. To decide which combination fits best the observed data we must carefully take into account the effect of the "fourth order correlation" f^{mnrs}. This hint at a "technical" detail may suffice here.

Once we have identified the v^{μ}s, we can perform the transformation (6.39) everywhere, i.e. in V (6.33) and in the constraints [cf. (6.29)–(6.32)]. The result is surprisingly simple. The transformed

$\tilde{V}\ (\xi)$ reads

$$\tilde{V}\ (\xi) = -\Lambda_0 - \sum_{\mu} C\lambda_{\mu}\xi_{\mu}^2 - \sum_{\mu} C_{\mu}\lambda_{\mu}\xi_{\mu}^4 + \Lambda'\left(\sum_{\mu}\xi_{\mu}^2\right)^2, \mu = 1, \ldots, L \tag{6.41}$$

The Lagrange parameters C, $C\mu$, Λ', Λ_0 are fixed by transformed constraints resulting from (6.29) to (6.32) and (6.34).

Summary of the learning procedure.

1. We evaluate (6.32) based on given data.
2. We seek the characteristic values λ_{μ} and characteristic vectors v^{μ} of the matrix (f^{mn}).
3. We perform the transformation (6.39) and (6.40) from the variables q_m to new variables ξ_{μ}.
4. The original potential V(q; Λ) (6.33) transforms into $\tilde{V}(\xi)$ (6.41). Because of its central role in the pattern recognition process, let us consider it in some detail. In order not to overload our discussion we consider the—in the present context-more relevant case of supervised learning. Then λ_{μ}s may be chosen such that all are equal,

$$\lambda_{\mu} = \lambda \tag{6.42}$$

We may treat Λ' as a free parameter. $\Lambda' > \max C_{\mu}$ for all μ [That can be achieved by a suitable choice of the parameter d in the constraint (6.31)]. While the coefficients C_{μ} can be calculated in principle, in many practical cases it suffices to assume all equal,

$$C_{\mu} = C' \tag{6.43}$$

Then we arrive at the fundamental potential function of the synergetic computer for pattern recognition that is of the form

$$\tilde{V}(\xi) = -\Lambda_0 - A\sum_{\mu}{}' \xi_{\mu}^2 - B\sum_{\mu}{}' \xi_{\mu}^4 + \Lambda'\left(\sum_{\mu} \xi_{\mu}^2\right)^2 \tag{6.44}$$

The sums $\sum_m'$ run only over those μ where $\lambda_{\mu} \neq 0$. As our analysis has shown, the valleys of $\tilde{V}(\xi)$, i.e., the attractors, are at the positions

$$\xi_K = \pm((\Lambda - B)/A)^{1/2}, \xi_{\mu} = 0 \text{ for } K \neq \mu \tag{6.45}$$

6.5.2 Pattern Recognition

A pattern q^0 to be recognized is decomposed into prototype patterns according to (6.39) so that the ξ_{μ}s are fixed. They mark a specific position in the potential landscape of $\tilde{V}(\xi)$, like a ball lying on a slope. From there it is pulled into the valley lying closest ("method of steepest descent"). If this valley is marked by *K*, then because of (6.45), the decomposition (6.39) reduces to

$$q_m = v_m^K const. \tag{6.46}$$

Which means that the pattern "*K*" has been recognized. So far, we presented the main steps of an algorithmic approach based on the variables ξ_{μ}. There is yet another approach that presumably comes close(r) to what is happening in the human brain. Namely consider the original potential V(q; Λ) (6.33). We may identify q_m with the activity of neuron "*m*", and the various Λs as synaptic strengths that are learned. The valleys of the potential landscapes V(q; Λ) are just the prototype patterns v^K. This claim can be directly substantiated by replacing $\xi\mu$, $\mu = 1, \ldots, L$ by means of the back transformation of (6.39), i.e.

$$\xi_{\mu} = \sum_m v_m^{\mu} q_m \tag{6.47}$$

and a concomitant transformation of the Lagrange multipliers (which means we can calculate them in (6.44) explicitly). A test pattern q_{test} is then "pulled" into the nearest valley ("steepest decent"). In mathematical terms this is achieved by solving the equation

$$\dot{q}(t) = -gradV$$

under the initial condition

$$q(t = 0) = q_{test}$$

6.6 Unbiased Modeling of Stochastic Processes Based on Observed Data

To elucidate our general approach (Haken 2006), we consider the time evolution of a single variable q that may be relevant for a complex system, e.g. the size of the population of a city. We assume that q is measured at times t_i, $i = 0, 1, \ldots, N$. In the spirit of Jaynes' principle, we wish to make an unbiased estimate on the joint probability distribution function.

$$P_N = P(q_N, t_N, q_{N-1}, t_{N-1}, \ldots, q_{0,}t_0) \tag{6.48}$$

To this end we maximize the information

$$S = -\int dV P \ln P \tag{6.49}$$

where the volume element is $dV = dq_0 dq_1 \ldots dq_N$ under given constraints. The limits of the multiple integrals are fixed by the available data. Our choice of constraints is facilitated if we assume that the process that leads to (6.48) is Markovian. This means that the "state" q of a system at time t_{i+1} is fixed *only* by its state at t_i. This allows us to split (6.48) into a product

$$P_N = \Pi_i P(q_{i+1}, t_i | q_i, t_i) P_0(q_{0,}t_o) \tag{6.50}$$

where the index i runs from 0 to $N - 1$. To simplify the notation we drop the times t_i so that the factors in (6.50) acquire the form

$$P(q_{i+1} | q_i) \tag{6.51}$$

Because we are interested in the temporal evolution of mean and variance we will use the constraints

$$f_{1,i} = \langle q_{i+1} \rangle_{q_i} \tag{6.52}$$

$$f_{2,i} = \langle q_{i+1}^2 \rangle_{q_i} \tag{6.53}$$

They allow us to relate our theoretical model (still to be derived) to observed data. In terms of theory, the averages $\langle ... \rangle$ are defined by

$$f_{k,i} = \int P(q_{i+1}|q_i) q_{i+1}^k dq_{i+1}, k = 1, 2 \tag{6.54}$$

But how to calculate them based on observed data? In a first step, let us recall the definition of the average of a random variable q. That we have denoted $\bar{q}$ [cf. (6.5), Sect. 6.1] and that we call now $\langle q \rangle$. Taking (6.5) and (6.1) together we may write

$$\langle q \rangle = \sum_q \frac{n(q)}{n} q, \sum_q n(q) = n \tag{6.55}$$

where $n(q)$ = number of cities with population size q lying within a interval of length $\Lambda[q - \Lambda/2, q + \Lambda/2]$. The whole q-axis is divided into such intervals. The choice of their length depends on the problem.

Now we may define

$$\langle q_{i+1} \rangle_{q_i} \tag{6.56}$$

by use of (6.55). In it, we just have to replace $n(q)$ by number of cities where number of citizens at time t_{i+1} is of size q_{i+1} (within interval Λ) *provided* its number at time t_i was q_i (within Λ). The average (6.53) can be calculated in any analogous fashion. After these preparations we apply Jaynes' principle to make an "unbiased guess" on (6.51) under the constraints (6.52)–(6.54) and normalization

$$\int P(q_{i+1}|q_i) dq_{i+1} = 1 \tag{6.57}$$

In all cases from now on we assume that the integration runs from $-\infty$ till $+\infty$. According to Sect. 6.3, $P(q_{i+1}|q_i)$ has the form

$$P(q_{i+1}|q_i) = \exp(\Lambda_0 + \Lambda_1 q_{i+1} + \Lambda_2 q_{i+1}^2) \tag{6.58}$$

The relations between the Λs and the contraints $f_{k,i}$ have been presented there; now we identify $f_{k,i}$ with f^k. Since $f_{k,i}$, $i = 1, 2$, are functions of q_i, so are the Λs. By means of elementary algebra and assuming $\Lambda_2 < 0$, we cast (6.58) into the form

$$\exp\left(-|\Lambda_2| \left(q_{i+1} - \frac{\Lambda_1}{2|\Lambda_2|} \right)^2 + \Lambda_0 + \frac{\Lambda_1^2}{4|\Lambda_2|} \right) \tag{6.59}$$

When we use (6.21b)–(6.21c) in (6.59) our original task is fulfilled. Now we want to go one step further. We assume that the times t_i are equally spaced,

$$t_{i+1} - t_i = \tau \tag{6.60}$$

and τ "sufficiently" small so that we may consider the limit $\tau \to 0$. We write $q_{i+\tau}$ instead of q_{i+1}. We require

$$\tau \to 0\colon P(q_{i+\tau}|q_\tau) \to \delta(q_{i+\tau} - q_i) \tag{6.61}$$

where δ is Dirac's δ-function. Because of (6.21b), (6.59) acquires the form

$$P(q_{i+\tau}|\mathrm{q_i}) = (|\Lambda_2(q_i)|/\pi)^{1/2}\exp\left(-|\Lambda_2|\left(q_{i+\tau} - \frac{\Lambda_1}{2|\Lambda_2|}\right)^2\right) \tag{6.62}$$

In order that (6.62) becomes a δ-function for $\tau \to 0$, we must require the behavior

$$|\Lambda_2| = \frac{W(q_i)}{\tau} \text{ for } \tau \to 0, \tag{6.63}$$

and (6.61) requires

$$\frac{\Lambda_1}{2|\Lambda_2|} \to q_i \text{ for } \tau \to 0 \tag{6.64}$$

For finite but small τ (6.64) generalizes to

$$\frac{\Lambda_1}{2|\Lambda_2|} = q_i + \tau K(q_i) \tag{6.65}$$

With (6.63) and (6.65) we obtain our final result of this section. In the short time limit, $\tau \to 0$, the conditional probability reads

$$P(q_{i+\tau}|q_i) = (W/(\tau\pi))^{1/2}\exp(-\frac{W}{\tau}(q_{i+\tau} - q_\tau - \tau K(q_i))^2) \tag{6.66}$$

where W may be a function of q_i. The result (6.66) allows us to derive a *Fokker–Planck—Ito* equation. Since the details are of a purely formal nature without giving us insight into any additional assumptions, we just present the final result. The probability distribution function P belonging to (6.66) for all i obeys

$$\dot{P}(q;t) = -\frac{\partial}{\partial q}(KP) + \frac{1}{4}W(q)^{-1}\frac{\partial^2 P}{\partial q^2} \tag{6.67}$$

When we just change the denotation $\frac{1}{4}W(q)^{-1} \to \frac{1}{2}G(q)$ we arrive at the Fokker–Planck-Ito equation of Chap. 5. For sake of completeness we mention two *generalizations*.

(1) Our procedure can be extended when we have to deal with a state vector $q = (q_1, \ldots, q_L)$.
In this case a Fokker–Planck-Ito equation of the form (5.51) with (5.52) results.
(2) All functions $N_j(q)$ *and* G_{jk} may be time-dependent.

Because of the one-to-one correspondence between Fokker–Planck equation and evolution equations (Chap. 5), the function N in (5.10) and (5.11) of Chap. 5 can be directly read off the drift-terms, i.e. K in (6.67). Similarly the strengths of the fluctuation forces in (6.9) and (6.9a) can be derived from G_{jk}.

6.7 Concluding Notes

In this Chap. 6 we have presented the 2nd Foundation of Synergetics—mathematical approaches that allow us to describe or predict states of complex systems based on a set of available data. Such approaches play a significant role in two interrelated contexts: Firstly, in the context of a researcher using partial/sparse/big data on the basis of which to simulate and understand aspects of urban dynamics. Secondly, as in Sect. 6.4, in the context of a newcomer to a city that has to take intuitive decisions and actions on the basis of the available partial/sparse/big data [but also on the basis of superfluous data (cf. the process of IA in Chap. 4)].

Our results have the form of a steady state or time dependent probability distributions of the relevant data/variables. In the context of urbanism, such data may be population size and total income. In the following Chap. 7, this explicit example will allow us to compare our approach with others having an equivalent goal. These are Baysian inference and Friston's free energy principle. Applications of our approach to urbanism will be presented in following chapters. Chapter 8 shows how it can be used to unearth interdependencies of the actions of citizens so to maintain a steady state. Our apporoach also allows us to study phase transitions of cities, e.g. caused by immigration waves (Chap. 10). The results of Sect. 6.6 form the basis of our treatment of the impact of the urban regulatory focus on the evolution of quiet slow-paced versus active fast-paced places (Chap. 13). Our approach of Sect. 6.5 to pattern recognition may find applications to urbanism in a variety of ways, e.g. the completion of city maps if only a small fraction of a city is known.

All in all we expect that the methodology developed in Chap. 6 will find numerous further applications in urbanism.

References

Bettencourt, L. M. A., Lobo, J., Helbing, D., Kühnert, C., & West, G. B. (2007). Growth, innovation, scaling, and the pace of life in cities. *Proceedings of National Academy Science U.S.A., 104*(17), 7301–7306. https://doi.org/10.1073/pnas.0610172104.

Bornstein, M. H., & Bornstein, H. G. (1976). The pace of life. *Nature, 1976*(259), 557–559.

Gigerenzer, G. (2015). *Simply rational. Decision making in the real world*. Oxford University Press.

Haken H. (2006) Information and Self-Organization: A Macroscopic Approach to Complex Systems, (3rd enlarged ed.). Springer.

Jaynes,E. T. (1957) Information theory and statistical mechanics. Physical Review 106(4) 620-650, Information theory and statistical mechanics II. Physical Review 108(2) 171-190

Levine, R., & Norenzayan, A. (1999). The pace of life in 31 countries. *Journal of Cross-Cultural Psychology, 30,* 178–205.

Milgram, S. (1970). The experience of living in cities. *Science, 167,* 1461–1468.

Walmsley, D. J., & Lewis, G. J. (1989). The pace of pedestrian flows in cities. *Environment Behavior, 21,* 123–150.

Weidlich, W. (2002). *Sociodynamics: A systematic approach to mathematical modelling in the social sciences*. London: Taylor & Francis.

Chapter 7
Relationships. Bayes, Friston, Jaynes and Synergetics 2nd Foundation

7.1 Introduction

In this chapter we want to elucidate relationships between Bayes' Theorem, Friston's Free Energy Principle (FEP), Jaynes' Maximum (Information) Entropy Principle, Synergetics' 2nd Foundation and the notion of SIRNIA (introduced in Chap. 4). We do this by means of examples from quantitative approaches to urbanism. The approaches we will present assume stationarity, i.e. the external and internal conditions don't change in the course of time.

Why do we want to elucidate the above relationships? Because there are several conceptual and mathematical similarities, as well as differences, between Friston's FEP, and the theoretical foundations of this study, namely, the general theory of synergetics (Chap. 3) and the notion of SIRNIA (Chap. 4) that was formulated in the context of *complexity theories of cities* (CTC) and the synergetic approach to cities. As we show below, by exploring the similarities and differences we shed light on, explicate and extend, several aspects of our theory.

The discussion below is developed in two steps: in the first (Sect. 7.1), we shortly introduce FEP and discuss the similarities and differences, while in the second (Sect. 7.2), we elucidate, as noted above the mathematical relationships between Bayes' Theorem, Friston's Free Energy Principle (FEP), Jaynes' Maximum (Information) Entropy Principle and Synergetics' 2nd Foundation. In the next Chap. 8 we further elucidate the relationship between the conceptual frameworks by focusing on the dynamics of steady states and the city.

H. Haken and J. Portugali, *Synergetic Cities: Information, Steady State and Phase Transition*, Springer Series in Synergetics,
https://doi.org/10.1007/978-3-030-63457-5_7

7.2 FEP, Synergetics and SIRNIA: Similarities and Differences

7.2.1 The Free Energy Principle

Referring to complex adaptive systems, specifically to biological agents such as animals or brains, "the free-energy principle, says that any selforganizing system that is at equilibrium with its environment must minimize its free energy … The principle is essentially a mathematical formulation of how adaptive systems resist a natural tendency to *disorder* … in the face of a constantly changing environment" (Friston (2010, 127). Referring to cognitive processes, the principle says "that biological agents must avoid *surprises* … [and, that] free energy is an upper bound on surprise, which means that if agents minimize free energy, they implicitly minimize surprise" (italics added).

The *free energy principle* originated from statistical mechanics by work of Feynman (1972) and Bogoliubov. When applied to life sciences by Friston in the context of neurodynamics and animal behavior (but also by other authors in other fields), FEP does no more refer to any kind of energy, but is a mathematical formalism to approximate probability distributions[1] based on an, in general, increasing number of measured data. The income distribution in a city is a simple example, that we interpret as probability distribution. First we devise a model, M, e.g. a Gaussian of the income variable q, but with its mean value as a free parameter f. FEP then allows one to calculate f with increasing precision. When two variables are involved, e.g. sensation, s, and action, q, (or as below, number of citizens and income), FEP allows us to calculate the probability p to measure a value q once we know s (or vice versa), minus $\log p$ is the surprise. Small surprise means large probability.

From the perspective of complexity theories, Friston's FEP refers, and adds insight, to the dynamics of steady states but not to that of phase transitions. It implies that CASs (complex adaptive systems) have innate tendency to avoid phase transition and perpetuate their steady state. An example suggested by Friston (2010) is to imagine a CAS which is a snowflake with wings (Fig. 7.1). Such a creature will restrict itself "to a domain of parameter space that is far from phase-boundaries… that would cause the snowflake to melt..", that is, to undergo phase transition . In a subsequent presentation, Friston (2011) further refers to Fig. 7.1 and asks: "what is the difference between a snowflake and a bird"? his answer: "… a bird can move (to avoid surprises)".

As is well recorded, birds (and other animals) move thousands of kilometers every year just in order to remain in the same ecological conditions. Like all life processes, such physical and cognitive (e.g. navigation) movements consume (free) energy. Yet the FEP does not refer to the role of energy in such processes, as noted.

[1] When such a probability distribution is written as an exponential function, the expression appearing in its exponent (multiplied by -1) is denoted Free Energy. It corresponds to the potential function V in Synergetics' 2nd foundation.

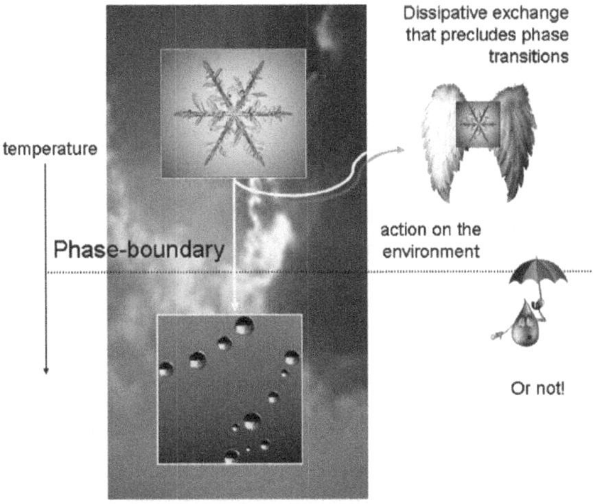

Fig. 7.1 "… By occupying a particular environmental niche, biological systems can restrict themselves to a domain of parameter space that is far from phase-boundaries. The phase-boundary depicted here is a temperature phase-boundary that would cause the snowflake to melt (i.e., induce a phase-transition) …". (Friston 2010, Fig. 1)

Rather, it is essentially a physicist's modeling approach and as such it doesn't get involved into the "energetic" details. The reason is that in contrast to physics, in life processes often there is a very complicated pathway from the elementary energy (ATP) processes to the final result, e.g. muscle contraction or sensation. As a model that employs mathematical tools from physics, FEP thus deals with relations between macroscopic features, e.g. perception and action, or as in our example below, number of citizens and income.

CASs thus restrict themselves to ecological environments ("niches") that enable them to maintain their steady state, while mentally to 'cognitive environments' that enable them to avoid surprise (that is, a cognitive phase transition). This, by means of an ongoing *perception–action* play between their *external states*—the data flow from the environment that generates sensory samples, and their *internal states*—the neural activity responsible to the recognition regarding the cause of a particular sensation (Fig. 7.2). These two states are separated by a *Markov blanket*—composed of sensory cells separating the states internal to the biological system from environmental states external to it.

7.2.2 FEP and Synergetics: Preliminary Comparison

Both conceptual frameworks theorize about complex, self-organizing systems. Friston's FEP focuses on CASs, specifically on the brain that is often described as 'the ultimate complex system'. Synergetics as a general theory of complex systems originated in physics, but later was intensively applied to CASs, cognition, brain dynamics and other domains, including cities and urbanism. SIRN, IA and their conjunction SIRNIA, were specifically developed in the context of complexity, cognition and the city (Portugali 2011; Portugali and Haken 2018).

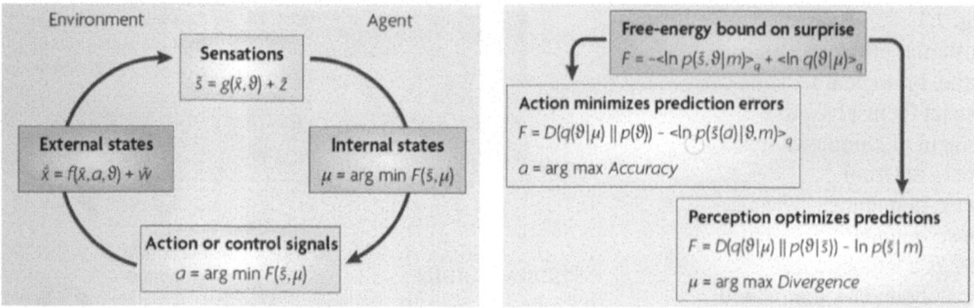

Fig. 7.2 Left: a scheme describing the quantities of free energy. Right: free energy can be reduced by means of action (selective sampling of sensory input) or by updating the internal model (Friston 2010)

Synergetics also influenced the FEP. "The central role of synergetics", writes Friston (2017, 143), "circular causality, separation of temporal scales and the slaving principle touches nearly every aspect of my field (neuroscience): from the action perception cycle through to radical enactivism". However, Synergetics and FEP differ in their focus of interest: FEP is specifically interested in the dynamics of steady states; synergetics as a general complexity theory deals with both steady state and phase transitions, yet (similarly to other complexity theories) with a clear inclination to, and interest in, the dynamics of phase transitions. The two conceptual frameworks in this respect thus complement each other.

The FEP is specifically akin to the 2nd foundation of synergetics (Chap. 6). According to both Friston's notion of "predictive coding", and to Synergetic's 2nd foundation, the brain is a kind of inference machine: following the last update (by action or sensation) it produces series of predictions (internal states) about the environment, examining and updating them by means of the information that comes from the environment (external states). In other words, top–down predictive models are compared with bottom–up representations by means of embodied action-perception. In this process, the prediction rule for the next step is based on the previously learned Free Energy, or, in terms of Synergetics' 2nd Foundation, on the potential landscape, V. FE and V may thus be interpreted as "generative models". The difference between Friston's FE and synergetics' V is: the learning of V is completed or supposedly completed whereas in FE it is going on.

To elucidate this procedure further, we consider the following scenario. In the dark, a rat is searching for food, e.g. a piece of cheese or meat. This "object" sends out smell molecules whose density decreases with increasing distance. In a first step, the rat turns its nose into an arbitrarily chosen direction, briefly sniffs, and registers the smell intensity stemming from the chosen direction. Then it repeats the same procedure with respect to another again arbitrary direction and compares the measured intensities. In the following step, it chooses preferentially a direction close to the former one with the higher intensity "under the hypothesis that the intensity doesn't decrease". If it does, prediction error is made, and the rat corrects its nose-direction accordingly. These steps are repeated again and again until a best fit is reached. In this way it minimizes surprise. Because of "noise", a perfect fit cannot be reached. This procedure requires that the object does not move. However, it also works if the object moves so slowly that there is enough time for the rat for a sufficiently good performance ("time-scale separation").

7.2.3 FEP and SIRNIA: A Comparison

There are also similarities between FEP and SIRNIA—both conceptualize a circularly causal play: FEP between internal *states* in the agents' mind/brain and external states referring to the environment; SIRNIA between internal *representations* constructed in the mind/brain of agents and external representations constructed in the environment. Yet, the above states and representations are not identical to each other. The internal representations are mind/brain constructs of the external world (e.g. a cognitive map), while external representations might take the form of agents' behaviour and action in the environment, but also stand-alone artificial objects the agents produce by means of their bodies, minds and tools (e.g. buildings, neighbourhoods and whole cities).

Both FEP and SIRNIA employ Shannon's (1948) theory of information (Shannon and Weaver 1949). Yet for FEP 'information' is a mathematical tool needed to convey the core task of FEP, which is: to minimize surprise in order to maintain the system in a steady state, away from phase transition. On the other hand, SIRNIA can be seen as a theory about the way agents process information when the task is to produce semantic information (e.g. pattern recognition) and pragmatic information (e.g. action/behaviour) by means of the inflation or deflation of Shannonian information; in some cases by maintaining the system in steady state, while in other cases (e.g. as in Fig. 7.3) by a sequence of phase transitions.

Similarly to FEP, in SIRNIA there is also an ongoing top–down–bottom–up interaction. Its IA canonical experiment is the process of vision as interpreted by Kandel, Hubel, Wiesel, and others (Fig. 4.5, Chap. 4). Here, data from the environment is being transformed into local Shannon information (SHI) by means of semantic information (SI). This local SHI triggers a top–down process that by means of I-inflation

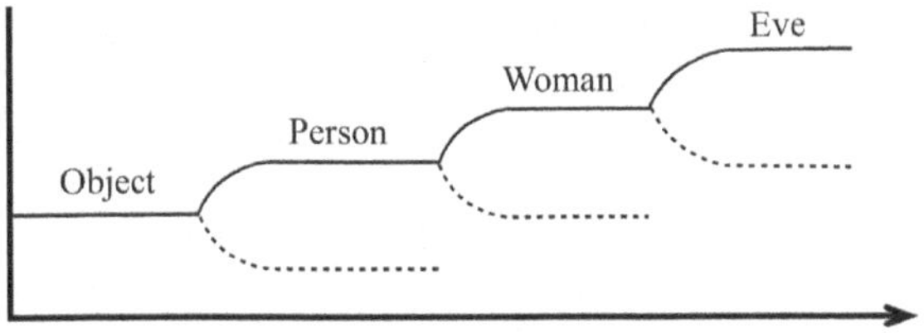

Fig. 7.3 A sequence of phase transitions in the approaching lady thought experiment (abscissa = amount of data vs. ordinate = recognized category/pattern). The broken line indicates the other dismissed options. For details see Haken and Portugali (2016)

or I-deflation, transforms the local SHI into global SI and pragmatic information (PI). Like in FE, in SIRNIA there is no independent bottom–up model of the environment that can be compared to a top–down predictive one. The result: we practically *see* a (Kaniza) triangle even though we *know* that this is a visual illusion—but we can and do resolve this discrepancy by our action, namely by looking at a finer scale. As a consequence, we are (even subconsciously) forced to invoke another previously learned "model." Our ambiguous interpretation of Arcimboldo's paintings (Fig. 7.4) is a typical example. Here attention plays a fundamental role (Köhler 1940; Ditzinger and Haken 1989, 1990). In FE, there may also be a set of data that can be better fitted by another model, though here there are no built in mechanisms similar to the attention approach. Quite generally, while in Friston's approach the task of the cognitive system is to minimize FE/surprise, in the SIRNIA conceptualization the task of the cognitive system is to produce meaning—abstract (SI) and action oriented (PI).

As can be seen, SIRNIA emphasizes the play between I-inflation/deflation in the process of perception—a play that is missing in FEP. On the other hand, what is missing in SIRNIA and is emphasized by FEP, is the role of feedback from the perception–action process: given a certain predictive perception (by means of I-inflation/deflation), the associated action feeds back by confirming or correcting the prediction. If it confirms, it reproduces and consolidates the model; otherwise, it

Fig. 7.4 A painting by Giuseppe Arcimboldo (1526–1593). At first sight, one recognizes a face. When one is looking at a finer scale, one recognizes just fruit and vegetables

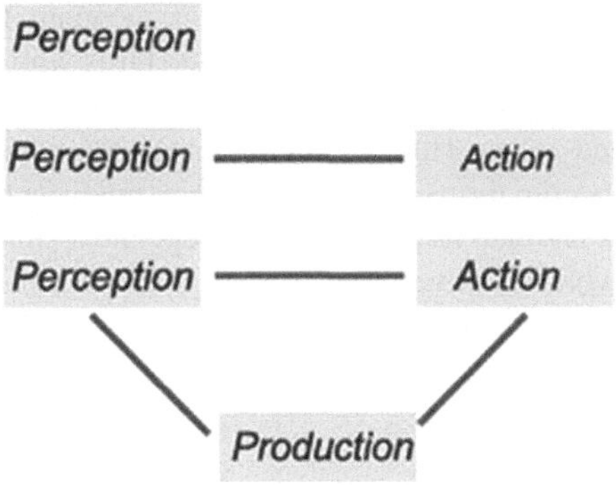

Fig. 7.5 The interrelationships between perception, action and production according to classical cognitivism (top), embodied cognition (center) and SIRN (bottom). In SIRN perception, action, and production form a single cognitive system (Portugali 2004, Fig. 3)

corrects and thus improves the model. A case in point is the popular party game: In the absence of a participant (the prospective searcher) an object is hidden. The searcher has to find it by his or her movements (action) in reaction to the participants' utterances "warm"/"cold" depending on the correct/incorrect (erroneous) step. In Friston´s terminology, each step in a direction that deviates, to the right or left, from the correct direction is an error. As can be seen from our presentation of Bayesian inference (Sect. 7.3.2 below), these errors compensate each other so that eventually the correct direction (in that Sect. 7.2.3 called "mean value") results. In the party game, this principle is verified by the "searcher" finding the hidden object. A second illustration of FE is a person going downhill in the dark, in which, in each new step the person tests the gradient (mathematically: method of stochastic steepest descent).

As just noted, FEP is intimately associated with the embodied, action-perception perspective on the process of cognition. SIRNIA with its play between internal and external representations adds a new dimension to that process (Fig. 7.5): the mind-brain-body is not only a kind of inference machine, but also a kind of 'construction or production machine'. It not only constructs an improved model of the environment, but by means of its capabilities for external representations it transforms the environment itself.

A case in point is the distinction between exploratory behavior and routinized behavior (Portugali and Haken 2018) revealed by ethological experiments of exploratory behavior (Golani et al. 1997, 1999): When an animal (e.g. a rat) is introduced to a new environment (an empty arena in the experiment), its first instinctive reaction is to explore it and to mark/construct in it stopping points of various sizes (measured by the time the rat spends there). This intensive but relatively short process continued until the whole arena is full of such points; then, the rat switches to routinized, slow, long-term and non-intensive behavior/activity which is much more relaxed and can be seen as a steady state activity and behavior. In other words, surprise reduction is implemented by the construction/production of an environment within which the animal can move freely without being surprised—similarly to Friston's imaginary creature (a snow flake with wings) in Fig. 7.1.

Animals' exploratory behavior is termed also *phenotypic behavior* (Golani et al. 1999) with the implication that such a behavior is programed by the long and slow process of Darwinian evolution; meaning, the rat has no choice but to construct the stopping points, similarly to the birds who have no choice but to build their nest in a specific form, the beavers that construct a dam, the termites that build "castles" and so on. Humans are different (Portugali 2011)—they have a choice as to the specific forms of the objects they produce; they thus produce *artifacts*, that is, "facts of art" that have their own independent *cultural evolution*—compared to Darwinian evolution, a very fast process that is intimately associated with brain' capacities of creativity, imagination, attention and feelings. The notion of SIRNIA was specifically designed to capture this process at an individual level as well as at the collective city (society) level.

7.3 Formalism

7.3.1 Some Basic Considerations

We start from what we may call "phenomenological urban allometry". In a first step we collect numerical data on an individual city, cities of a whole country or an ensemble of cities across several counties. In view of the high mobility of citizens, the just mentioned stationarity condition may pose a problem. For instance in cities such as Tel Aviv the number of people living there at day and at night differs by several hundred thousand. So how to secure stationarity? At least in principle, one can measure the "population" size every—say—two hours. Then "stationarity" refers to the *same* time at *all* days and nights. Having this refinement in mind, we turn to the general case.

The data may be the number of citizens s ("size" of a city), quantities q such as the total income, number of cars, etc. In the spirit of allometry we seek simple relations ("laws") between these "indicators", e.g. total income q and number of citizens s. The simplest relation is proportionality.

$$q = rs \tag{7.1}$$

Where r is a constant ("ratio"). Our goal is to verify the law (7.1) and to determine r by means of a series of "measurements", i.e. polls/ sampling of pairs of s and q. In other words, we are looking for a correlation between s and q. Let the results be (q_1, s_1) and (q_2, s_2), ... Then we may form

$$\frac{q_n}{s_n} = r_n, n = 1, \ldots, N \tag{7.2}$$

and determine the best fit of r by means of

$$r = \frac{1}{N}\sum_{n=1}^{N} r_n,\ N \to \infty \tag{7.3}$$

Another approach to determine r is to use in (7.1) just the mean values $\langle s \rangle$, $\langle q \rangle$ (cf. Chap. 6).

The approaches we are going to deal with in this chapter are more ambitious. For instance, when we deal with ensembles of cities, we have to consider small, medium size, and large cities. So the use of mean values only may not be sufficient. So, quite often one goes one step further by taking the variance S, or equivalently, the standard derivation σ into account, where

$$S = \langle q^2 \rangle - \langle q \rangle^2 \tag{7.4}$$

$$\sigma = S^{1/2} \tag{7.5}$$

How can we still verify a law of the form (7.1) and the numerical value of r based on measurement of s and q? To this end we employ an approach developed by Synergetics' 2nd foundation. In a first step we search for appropriate average values such as

$$\langle s \rangle, \langle s^2 \rangle, \langle q \rangle, \langle q^2 \rangle \tag{7.6}$$

that may be obtained by some sampling ("sparse data"). Because we hypothesize (7.1) that implies a *correlation* between q and s, we seek the simplest correlation function between q and s, namely

$$\langle qs \rangle \tag{7.7}$$

It may be related to (7.1) by forming

$$q - rs \tag{7.7a}$$

and taking the average over the square

$$\langle (q - rs)^2 \rangle = \langle q^2 \rangle - 2r\langle qs \rangle + \langle s^2 \rangle \tag{7.7b}$$

Now we have the complete set (7.6)–(7.7) of the *constraints* to apply Jaynes' MIE principle to derive the best probability distribution $p(q, s)$.

According to Chap. 6, Sect. 6.3, it reads

$$p(q, s) = \exp(\Lambda_0 + \Lambda_1 q + \Lambda_2 q^2 + \Lambda_3 qs + \Lambda_4 s + \Lambda_s s^2) \tag{7.8}$$

To determine the Lagrange parameters Λ we insert (7.8) in (7.6) and (7.7) which leads us to equations that generalize (6.21a) of Chap. 6. Furthermore, we take care of the normalization condition (6.20) of Chap. 6. We use the obvious notation

$$\langle q^2 \rangle = f_{qq}, \langle q \rangle = f_q, \langle qs \rangle = f_{qs}, \langle s \rangle = f_s, \langle s^2 \rangle = f_{ss} \tag{7.9}$$

At first, we check whether q and s are correlated at all. The criterion reads

$$\langle qs \rangle - \langle q \rangle \langle s \rangle \neq 0 \tag{7.9a}$$

We assume that this relation is fulfilled. Because of the use of a Gaussian (7.8), all average values can be calculated explicitly and expressed as functions of the Λs, e.g.

$$f_q = f_q(\Lambda_0, \Lambda_1, \ldots, \Lambda_s) \tag{7.8a}$$

Thus we obtain a set of equations based on (7.8), where the l.h.s, is given experimentally (cf. Chap. 6, Sect. 6.2).

These equations allow us to obtain expressions for the Λs as functions of the observed averages (7.6) and (7.7), see below.

To bring out the essentials and to prepare the ground for our subsequent comparison with Bayes' Theorem and Friston's Free Energy Principle we assume $\Lambda_2 = -1$, $\Lambda_1 = 0$ and use a slightly simpler notation.

$\Lambda_4 = 2s_0$, $\Lambda_5 = -1$ so that (7.8) acquires the form

$$p(q, s) = C\exp(-q^2 + 2\beta qs + 2s_o s - s^2) \tag{7.10}$$

where C is the normalization constant. By means of (7.10) we find the announced explicit expressions for (7.6) e.g.

$$\langle s \rangle = \frac{s_o}{1-\beta^2}, \langle q \rangle = \beta \frac{s_o}{1-\beta^2} \tag{7.10a}$$

$$\langle qs \rangle - \langle q \rangle \langle s \rangle = \frac{\beta}{1-\beta^2} \tag{7.10b}$$

The l.h.s. of (7.10 a, b) are to be interpreted as observed/ measured. In accordance with (7.9a), (7.10b) allows us to check whether there is a correlation between the "systems" q and s. From (7.10a) we deduce the relation.

$$\langle q \rangle = \beta \langle s \rangle \tag{7.10c}$$

By comparing it with (7.1), we can calculate $r = \beta$ from observed data.

There is a second way to calculate β (or r). We seek the maximum of (7.10) or the minimum of the exponent in (7.10). We obtain by taking the derivatives (and dividing by 2)

$$-q + \beta s = 0 \tag{7.11}$$

$$-s + s_o + \beta q = 0 \tag{7.12}$$

We obtain ("m" = maximum)

$$q_m = \beta s_m \tag{7.13}$$

$$s_m = s_o(1 - \beta^2)^{-1} \tag{7.14}$$

By comparing (7.13) with (7.1) we obtain our essential result

$$r = \beta \tag{7.15}$$

To understand the meaning of s_m (7.14) we invoke the explicit form of $p(s)$ that is defined by $\int\limits_{-\infty}^{\infty} p(q, s)dq = p(s)$.

By use of (7.10) we obtain

$$p(s) = \text{const.exp}(-(1 - \beta^2)s^2 + 2ss_0) \tag{7.16}$$

Obviously the maximum of $p(s)$ lies at $s = s_m$ (7.14). According to (7.16) our approach is only valid if $\beta^2 < 1$ because otherwise (7.16) diverges for $|s| \to \infty$. We may derive $p(q)$ in an analogous way. We then observe that $\beta^2 < 1$ to guarantee the existence of p(q)—otherwise we must re-scale q or s. (use of other units). We may make prediction on the outcome of a further measurement on q once we have measured s finding $s = s_1$. To use this we write (7.10) as (note $\beta = r$!)

$$p(q, s_1) = C'\exp\left(-(q - rs_1)^2\right)\exp\left(-\left(1 - r^2\right)s_1^2 + 2s_o s_1\right) \tag{7.17}$$

where the normalization constant C′ can be split into two factors belonging to the first and second exp-function, respectively.

$$C' = \frac{1}{\sqrt{\pi}}\left(\frac{1 - r^2}{\pi}\right)^{\frac{1}{2}} \exp\left(-s_o^2\left(1 - r^2\right)^{-1}\right) \tag{7.18}$$

The normalized first exp-function represents the conditional probability

$$p(q|s_1) = \frac{1}{\sqrt{\pi}}\exp(-(q - rs_1)^2) \tag{7.19}$$

Equation (7.19) represents the probability to measure a value $q = q_1$ provided s_1 has been measured.

In the exponent, $(q - rs_1)^2$ is an explicit example of "surprise". It is minimal, if $q = rs_1$.

A final "technical" note:

Our "minimal" model captures the essential features of the correlation between q and s. We can cover the general case by the substitution $s \to \sqrt{\gamma} s$, $q \to \sqrt{\alpha}(q\text{-}q_o)$, so that we have the five parameters α, q_o, β, s_o, γ corresponding to $\Lambda_1 - \Lambda_5$ to satisfy the five constraints (7.9).

Note that Λ_o is fixed by the normalization condition.

7.3.2 Bayes' Theorem Scrutinized

According to Chap. 6, Sect. 6.3, the best unbiased guess on a probability distribution, say $p(q)$, if only mean $\langle q \rangle$ and variance S are known, is a Gaussian that we write as

$$p(q) = \sqrt{\frac{\alpha}{\pi}} \exp(-\alpha(q - q_o)^2) \tag{7.21}$$

where $q_o = \langle q \rangle$, and $\sqrt{\alpha/\pi}$ is a normalization factor. For sake of completeness we quote $p(s)$,

$$p(s) = \sqrt{\frac{\alpha}{\pi}} \exp(-\gamma(s - s_o)^2) \tag{7.22}$$

Now the crucial question arises:

While according to (7.1), now written as [cf. (7.7a, b)]

$$\langle (q - rs)^2 \rangle = 0 \tag{7.23}$$

there exists relation between q and s, there is no obvious relation between (7.21) and (7.22). How can we find the missing link?

The key to find that link is provided by our derivation of Bayes' Theorem in Chap. 6, Sect. 6.1. There we started from the joint probability $p(m, n)$ which in our present variables q, s reads $p(q, s)$.

When we integrate over s, we obtain the probability distribution $p(q)$ above. According to (7.21), $p(q)$ should be a Gaussian. In complete analogy, also $p(s)$ should be a Gaussian. For sake of clarity we treat the special case, $\alpha = 1$, $\gamma = 1$, $q_o = 0$.

Then we expect $p(q, s)$ to be of the form

$$p(q, s) = N \exp(-q^2 - (s - s_o)^2 + ?) \tag{7.24}$$

N is the normalization constant. The question mark "?" stands for the missing link.

The only term that fits into the quadratic form in the exponent in (7.24) is

$$2\beta qs \tag{7.25}$$

where β is a parameter still to be fixed by a series of measurements (cf. Sect. 7.1) and the factor 2 has been added to make the following somewhat simpler. Now we do the decisive step. Namely the resulting quadratic form

$$q^2 + (s - s_o)^2 - 2\beta qs \tag{7.26}$$

can be decomposed in two ways

(a)

$$(q - \beta s)^2 - \beta^2 s^2 + (s - s_o)^2 \tag{7.27a}$$

and

(b)

$$q^2(1 - \beta^2) - 2\beta s_0 q + (s - s_0 - \beta q)^2 \tag{7.27b}$$

When we lift (7.27a) to the exponent [cf. (7.24)] (with minus sign), we arrive at

$$\begin{array}{ccc} p(q,s) = N_1 \exp(-(q-\beta s)^2) \cdot & N_2 exp(-(1-\beta^2)s^2 + 2ss_o - s_o^2) & \\ \downarrow & \downarrow & (7.28a) \\ p(q|s) & p(s) & \end{array}$$

i.e. joint probability = conditional probability times single probability.

In complete analogy we obtain

$$\begin{array}{ccc} p(q,s) = N_1' \exp(-(1-\beta^2)q^2 - 2\beta s_0 q) \cdot & N_2' \exp((s - s_0 - \beta q)^2) & \\ \downarrow & \downarrow & (7.28b) \\ p(q) & p(s|q) & \end{array}$$

In (7.28a, b) the normalization factor N has been split into two appropriate factors. Now we can do the last step à la Bayes. We equate (7.28a, b) and divide both sides by $p(s)$ (we add the normalization factors).

$$\frac{1}{\sqrt{\pi}}\exp(-(q-\beta s)^2)=\frac{1}{\sqrt{\pi}}\exp(-(s-s_0-\beta q)^2)\cdot\frac{p(q)}{p(s)} \tag{7.29}$$

Equation (7.29) provides us with an explicit example of Bayes' Theorem. In Sect. 7.3.3 we present an application.

7.3.3 Bayesian Inference—An Example

By which probability distribution $p(q|\beta)$ can we reproduce the mean value

$$\langle q\rangle = q_m? \tag{7.30}$$

We define a "model"

$$p(q|\beta)=e^{-(q-\beta)^2} \tag{7.31}$$

And multiply it by the "prior" with variable β

$$e^{-\beta^2} \tag{7.32}$$

Under the 'belief" that $\beta=0$ in (7.31) is most probable (we omit normalization factors).

The product between (7.31) and (7.32),

$$p(q|\beta)p(\beta)=e^{-(q-\beta)^2}e^{-\beta^2} \tag{7.33}$$

can be written as

$$\begin{matrix} e^{-2(\beta-\frac{q}{2})^2} & e^{-\frac{1}{2}q^2} \\ \downarrow & \downarrow \\ p(\beta|q) & p(q) \end{matrix} \tag{7.34}$$

In the spirit of Bayes' Theorem we equate (7.33) and (7.34) and rearrange

$$e^{-2(\beta-\frac{q}{2})^2}=e^{-(q-\beta)^2}e^{-\beta^2}p(q)^{-1} \tag{7.35}$$

where the l.h.s. is the "posterior", and (7.35) is valid up to a normalization constant.

In view of actually measured values of q giving rise to (7.30) the relation (7.35) may only be considered as a first step of an approximation procedure where we improve the prior stepwise. This is done by using at each new step the old posterior as new prior (under the improved "belief"). We assume the prior of the nth steps in the form (up to normalization)

$$p_n(\beta) = e^{-n(\beta-\beta_n)^2} \tag{7.36}$$

We use it on the r.h.s. of (7.35) jointly with a newly measured value of $q = q_n$.

$$e^{-(q_n-\beta)^2} e^{-n(\beta-\beta_n)^2} p_n(q_n)^{-1} \tag{7.37}$$

Which can be rearranged to

$$e^{-(n+1)\left(\beta-\frac{n}{n+1}\beta_n-\frac{1}{n+1}q_n\right)^2} e^{f(n,q_n)} \tag{7.38}$$

The first factor represents the posterior of the $n+1$ step, that according to (7.36) must read

$$p_{n+1}(\beta) = e^{-(n+1)(\beta-\beta_{n+1})^2} \tag{7.39}$$

Comparing (7.39) with (7.38) we obtain the recursion relation

$$\beta_{n+1} = \frac{n}{n+1}\beta_n + \frac{1}{n+1}q_{n+1} \tag{7.40}$$

Its solution with $\beta_0 = q_0 = 0$ reads

$$\beta_{n+1} = \frac{1}{n+1}\sum_{k=0}^{n} q_k \tag{7.41}$$

For $n \to \infty$, β_{n+1} coincides with the definition of (7.30). Note that (7.36) with its proper normalization factor $\sqrt{\frac{n}{\pi}}$ yields for $n \to \infty$ just Dirac's δ-function, i.e. β acquires a fixed value.

Our simple example is quite representative for Bayesian inference. First a model, in many particular cases, a Gaussian of several variables and with free parameters is formulated. Then, by use of measured data and Bayes' Theorem, the parameters are calculated.

7.3.4 Friston's Free Energy Principle

In this Section we present an explicit example of Friston's Free Energy Principle and compare it with the approach of Sect. 7.3.1. We start from a "generative model" with three variables q, s, ψ and the parameters a, s_0, β,

$$\check{V}(q, s, \psi) = q^2 - 2\beta qs + s^2 - 2s\psi + a(\psi - s_0)^2 \tag{7.42}$$

More precisely, the generative model sensu Friston does not contain the last term containing a that stems from the prior (cf. previous section). Here we anticipate the result of the following Sect. 7.3.5, where by use of Bayesian inference we determine a and s_o so that we may directly start from (7.42). The still free parameter β is finally fixed by minimizing (7.44).

The joint probability distribution reads

$$\check{P}(q, s, \psi) = \check{N}\exp\Big(-\check{V}(q, s, \psi)\Big) \tag{7.43}$$

$\check{N}$: Normalization constant. The variables run from $-\infty$ to ∞.

We calculate Friston's Free Energy using his definition (2013). To this end, we need

$$p(\psi, s),\ p(\psi|s),\ p(s),\ p(\psi|q).$$

These probabilities can be calculated by means of (7.43). We obtain $p(\psi, s)$ by integrating (7.43) over q and $p(s)$ by integrating $p(\psi, s)$ over s. We derive the conditional probability by means of $p(\psi, s) = p(\psi|s)p(s)$. We derive $p(\psi|q)$ in an analogous way. In all these cases we obtain Gaussian distributions. Now we apply Friston's definition of his Free Energy

$$F(s, q) = D(p(\psi|q)/p(\psi|s)) - \ln p(s) \tag{7.44}$$

where D in the Kullback-Leibler distance

$$D = \int_{-\infty}^{\infty} p(\psi|q)\ln(p(\psi|q)/p(\psi|s))d\psi \tag{7.45}$$

Because of the Gaussians, the integration can easily be performed. For an intended comparison with Sect. 7.3.2 it suffices to consider the limiting case $a \gg 1$. Then

$$F(s, q) = Const. + \frac{1}{a}(s_0 + \beta q - s)^2 + (1 - \beta^2)(s - \frac{s_0}{1 - \beta^2})^2 \tag{7.46}$$

For a comparison with Sect. 7.3.1, i.e Synergetics 2nd Foundation, we again start from (7.43) from which we eliminate ψ

$$p(q, s) = \int_{-\infty}^{\infty} d\psi \, \check{N} \exp\left(-\check{V}\right) = Nexp(-V(q, s)) \tag{7.47}$$

where

$$N = \pi^{-1}(1 - \beta^2 - \frac{1}{a})^{1/2} \tag{7.48}$$

and, for $a \gg 1$

$$V(q, s) = q^2 + s^2 - 2qs\beta - 2s_0 s + \frac{s_0^2}{1 - \beta^2} \tag{7.49}$$

which is the synergetic potential.

This potential is fixed up to an additive constant. In (7.49) we add and subtract $\beta^2 s^2$ to obtain

$$V(q, s) = (q - \beta s)^2 + (1 - \beta^2)(s - \frac{s_0}{1 - \beta^2})^2 \tag{7.50}$$

Now we are in a position to discuss commonalities and differences between $F(7.46)$ and $V(7.50)$. F and V are the sums of two quadratic terms, where the second terms coincide, whereas the first terms depend differently on q and s. Nevertheless, the unique *minima* of F and V coincide.

Minima: F: $(s_0 + \beta q - s) = 0$, $s - \frac{s_0}{1-\beta^2} = 0$

From which follows $q - \beta s = 0$

Equivalently, V: $q - \beta s = 0$ $s - \frac{s_0}{1-\beta^2} = 0$

By means of the minima, we may either calculate s_0, β given the mean values of q and s or vice versa.

7.3.5 *Prospective Processing*

We try to follow Friston's "philosophy" as closely as possible.

Again we consider a system with only two variables q, s. Friston's explicit example identifies s with the state variable of a sensory neuron that is coupled bidirectionally to a neuron with state variable q that, e.g., may rule action. The sensory neuron receives signals of size ψ from a "hidden" source. How can the sensory neuron learn the statistical distribution of ψ, $p(\psi)$? To this end we define a "generative model"

$$q^2 - 2\beta qs + s^2 - 2s\psi \tag{7.51}$$

We assume that β is a fixed parameter.

To derive $p(\psi)$ from observed data on q and s, we apply the method of Bayesian inference, Sect. 7.3.3. In slight generalization of Sect. 7.3.3 we form the joint probability as

$$\exp(-q^2 + 2\beta qs - s^2 + 2\psi s)p(\psi) \tag{7.52}$$

where we calculate the prior $p(\psi)$ iteratively.

According to Sect. 7.3.3, after the n'th iteration, $p(\psi)$ must be of the form

$$p_n(\psi) = Ne^{-a_n(\psi-\psi_n)^2} \tag{7.53}$$

where $a_n \to \infty$ with $n \to \infty$ and ψ_n depends on all the previously measured pairs

$$q_l, s_l, l = 0, \ldots, n-1$$

Now let us compare (7.52) and (7.53) with (7.42) and (7.43). Quite obviously a_n corresponds to a, and s_0 to ψ_n. This means that the variance of ψ shrinks—the predicted values ψ become more and more precise as well as s_0 (alias as ψ_n). But the essential conclusion is that the probabilistic prediction on q, s, becomes more and more precise with more and more collected data on q, s.

Clearly, the interpretation of q, s, as state variables of neurons is irrelevant to urbanism. But we may interpret q and s as urban indicators on which the urban agent collects data. Or the agent reacts, by a specific action (quantified by q), to an indicator s. The more precise the information on s, the better the prediction on the action of the agent. This is part of the "mechanism" which stabilizes urban life. This stabilization effect is also reflected by the minimization of Friston's Free Energy.

7.3.6 Comparison

In Sect. 7.3.1 we apply Jaynes' principle to calculate the joint probability of q, s, based on the measurements of mean, variance and correlations between q, s. The implicit assumption is made that the measurements are sufficient. In this sense, we obtain an explicit and "exact" potential function $V(q, s)$. Both Friston's Free Energy approach and the related predictive processing make the approximate character of the measurements/data collection explicit (our parameter a). in the limit of sufficient measurements, $a \to \infty$, the minima of V and F coincide. But while in V q and s play a symmetric role, in F the role of s is prominent, especially if $a \to \infty$.

7.4 Summary. A Bird's Eye View

In view of their potential applications to "theoretical urbanism" we have elucidated several approaches, namely (a) "Bayes/Friston" and (b) "Jaynes/Synergetics 2" by means of simple examples. Both (a) and (b) aim at the derivation of a probability distribution of appropriate variables (in urbanism indicators) in accordance with measured data. In a first step, both approaches (a), (b) require the formulation of a "generative model". At least in principle in (a) its choice is left to the modeler; though in numerous practical cases the choice is "Gaussians". But there are approaches called "Variational Bayesian" that go further and are mathematically demanding. In the approach (b) the selection is implicitly made by the choice of moments as constraints. While moments up to second order give rise to Gaussians, in some cases, e.g. pattern recognition, Chap. 6, fourth order terms have been included. Both (a) and (b) contain free parameters. Their calculation is a central task.

Bayesian inference puts in evidence how the precision of parameter determination is increased with an increasing number of measurements. Thus Bayesian inference and its further development by Friston´s FEP approach shed light on the learning process based on the sensation/action cycle. It is here where the fundamental difference between (a) and (b) comes to light. In (b) it is assumed that the measured data (even sparse data) suffice to calculate the average moments needed as constraints. Thus in contrast to (a), the approach (b) refrains from this "quality check". On the other hand this approach is more straightforward.

References

Ditzinger, T., & Haken, H. (1989). Oscillations in the perception of ambiguous patterns. *Biology Cybernetics, 61*, 279–287.

Ditzinger, T., & Haken, H. (1990). The impact of flucuations on the recognition of ambiguous patterns. *Biology Cybernetics, 63*, 453–456.

Friston, K. (2010). The free-energy principle: A unified brain theory? *National Review Neuroscience, 11*, 127–138. https://doi.org/10.1038/nrn2787.

Feynman, R. A. (1972). *Statistical mechanics*. MA Benjamin: Readings.

Friston, K. (2011). Functional and effective connectivity: A review. *Brain Connectivity, 1*, 1. Published Online: 1 Jun 2011. https://doi.org/10.1089/brain.2011.0008.

Friston, K. (2017). Epistemic patterns—A tribute to Hermann Haken. In J. Kriz & W. Tschacher (Eds.), *Synergetik als Ordner - Die strukturierende Wirkung der interdisziplinären Ideen Hermann Hakens* (pp. 143–146). Lengerich: PABST Science Publishers.

Golani, I., Einat, C., Tchernichovsky, O., & Teitelboum, P. (1997). Keeping the body straight in the locomotion of normal and dopamine stimulant treated rats. *Journal of Motor Behavior, 29*, 99–112.

Golani, I., Kafkafi, N., & Drai, D. (1999). Phenotyping stereotypic behaviour: Collective variables, range of variation and predictability. *Applied Animal Behaviour Science, 65,* 191–220.
Haken, H., & Portugali, J. (2016). Information and selforganization. A unifying approach and applications. *Entropy, 18,* 197. https://doi.org/10.3390/e18060197
Köhler, W. (1940). *Dynamics in psychology*. N.Y.: Liveright.
Portugali, J. (2004). Toward a cognitive approach to urban modelling. *Environmental Planning B, 31,* 589–613.
Portugali, J. (2011). *Complexity, cognition and the city*. Berlin/Heidelberg/New York: Springer.
Portugali, J., & Haken, H. (2018). Movement, cognition and the city. In J. Portugali (Ed.), *Cognition and the city. Built environment* (Vol. 44, No. 2, pp. 136–161). Alexandrina Press.
Shannon, C. E. (1948). A mathematical theory of communication. *Bell System Technical Journal, 27*(379–423), 623–656.
Shannon, C. E., & Weaver, W. (1949). *The mathematical theory of communication*. Illinois: Univ of Illinois Press.

Part II
Steady State and Phase Transition

Chapter 8
Steady States and the City

8.1 Introduction

8.1.1 General

As elaborated above (Chap. 2), the last decades have witnessed the emergence of *complexity theories of cities* (CTC)—a domain of research that applies the various theories of complexity to the study of cities. Two basic concepts in complexity theories are *steady state* (StS)—the process by which the system maintains its order in face of perturbations/fluctuations, and *phase transition* (PT)—the process by which a system undergoes a qualitative change. Together with other notions such as emergence, self-organization, bottom–up, fractal structure and the like, StS and PT are part and parcel of the language and conceptual framework of complexity theories. And yet, looking at the evolving discourse in the domain of CTC, it can be observed that compared to the other CTC notions, StS and PT have received only marginal attention. It can further be observed, that when StS and PT are being referred to in CTC, they are presented in the most general way ignoring the fact that there are several forms of StS and PT.

Our view is that this is due to the specific history of the study of cities—the fact that it underwent two paradigmatic revolutions (e.g. phase transitions), that resemble Snow's (1964) *Two Cultures*: one in the early 1950s: the so called *quantitative revolution* that attempted to transform the *study* of cities into a *science*, and another since the early 1970s that criticized the first and applied social theory structuralist-Marxist-Humanistic (SMH) perspectives to the study of cities. While for a few of its practitioners CTC was seen as an opportunity to bridge the gap between the two cultures of cities (e.g. Portugali 2000, 2011), for the majority it was seen as an opportunity to revive the quantitative approach to cities on the sounder bases of complexity theory as *The New Science of Cities* (Batty 2013). The result was a kind of division of labor, where most CTC studies focus on data-rich, short-term, more technical

H. Haken and J. Portugali, *Synergetic Cities: Information, Steady State and Phase Transition*, Springer Series in Synergetics,
https://doi.org/10.1007/978-3-030-63457-5_8

topics that afford quantification, leaving the qualitative long-term aspects of urban dynamics, to students of SMH.

A recent case of data-rich topic that affords quantification is *urban allometry* (or *scaling*) that commences from the view "that important demographic, socioeconomic, and behavioral urban indicators are, on average, scaling functions of city size that are quantitatively consistent across different nations and times" (Bettencourt et al. 2007, pp. 7301–2; Bettencourt 2013). Following two recent studies in which we explore the various aspects of urban allometry studies (cf. Chaps. 12 and 13 below), we realized that the different perspectives on urban allometry are a consequence of different assumptions regarding the qualitative StS and PT processes that underlie urban change. The problem is, however, that these assumptions are *implicit* and as such are not spelled out. There is thus a need to explicate the role of StS and PT in the dynamics of cities and this is exactly the aim of the present and next chapters—in the present chapter, the role of StS, while in Chaps. 9 and 10 that follow, the role of PT.

8.1.2 Stasis is Data

Stasis is data is the motto of Eldredge and Gould's (1972) famous *punctuated equilibrium (PE)*—a notion that has turned "… the basic fact of paleontology from an unstated embarrassment into a subject of active and burgeoning research." (Gould 1991). This is surprising since the notion punctuated equilibrium focuses on the dynamics and form of change and yet Gould says that its significance is in directing attention to *stasis,* that is, to no change. Previous to PE, writes Gould, it was common to see "evolution as gradual change, the stability of species counted as no data—that is, as absence of evolution […] Punctuated equilibrium has changed the context. Stasis has become interesting as a central prediction of our theory. […] Don't we want to know why so many species *don't* change for so long? Stasis is a puzzle, not a negativity."

In the language of complexity theory, "stasis" is 'steady state' while punctuated equilibrium 'phase transition'. From this perspective, the motto ***stasis is data*** reads "steady state is data" and its significance is in the question 'what is the dynamics of a steady state, how is it that a system that is always far from equilibrium, always changing and never in rest, manages to maintain its structure?' The question is relevant since most attention in the various theories of complexity is directed toward change. This is so in general and this is specifically so in the domain of CTC.

It is interesting to note that the question of 'stasis' is central to the domain of social theory that dominates the study of cities since the early 1970s. The basic question: How is it that social structures and socio-spatial structures (e.g. cities) maintain their structure despite basic internal contradictions, conflicts and inconsistencies, between rich versus poor, free-versus slave persons … etc.? The basic answer: by means of *social reproduction*: the social structure (OP in the language of synergetics) describes and prescribes the behavior of the citizens, that by means of their action strengthen, support and *perpetuate* the social structure and so on. In the parlance of synergetics one would say "and so on in *circular causality*", that is, social reproduction in social theory is circular causality in synergetics. Social theory further specifies how the social structure (OP in the language of synergetics) prescribes the behavior and action of the citizens: by means of its institutions: government, police, law, and by means of the prevailing social ideology that legitimizes the existing (often contradictory) social relations. E.g. think about slavery.

The most prominent expression for this process of circular causality is *daily life* and daily *routines* (or rhythms) as studied by several social theorists such as Goffman (1956), Lefebvre (1968/1994), Giddens (1979, 1984), de Certeau (1984) and others. And this takes us to our previous discussion in Chap. 7 that explored Friston's FEP in relations to Synergetics. As we've seen there, the FEP and its synergetic's parallels are essentially theories of systems in steady state. Our aim in this chapter is thus to complement and extend the discussion on StS (in Chaps. 9 and 10 that follow, we shall focus on PT). As in previous chapters, our main focus of interest would be the city and its dynamics which we will examine from the perspective of Synergetics (Chaps. 3, 5 and 6) and its two associated notions—SIRN, IA and their conjunction SIRNIA (Chap. 4).

8.1.3 Flux Equilibrium

There is a rich body of literature in archaeology, anthropology, sociology, social philosophy, architecture, geography and CTC too, on the origin of urbanism and subsequent processes of urbanism (Portugali 2020/1 and further bibliography there). Common to most is, firstly, that they focus on phase transitions ("urban revolutions") but tend to ignore the steady state dynamics. Secondly, that they assume that the structure and dynamics of settlements and cities, with their order and stability, must have come into existence by some kind of phase transitions like species in evolution. On an abstract level we may think of a "Darwinian" competition between order parameters each governing a specific realization. Like in biology (why this species and not another one ?) the basic processes may have been an interplay between chance events, external conditions and said competition, but also cultural inheritance.

Let us turn to the eventually established steady state. On the face of it, its macroscopic features are time-independent. We recognize static structures such as streets, buildings, infrastructures, artifacts, but also laws, regulations, etc.. Yet they are only relatively time-independent. In a paper from 1999, Weidlich referred to the play between slow (e.g. a regional system) and fast (e.g. a city) variables/systems, when the slow can be treated as static variables and as such provide boundary conditions (order parameters or control parameters) to the faster ones.

"No man ever steps in the same river twice, for it's not the same river and he's not the same man", said Heraclitus—the pre-Socratic Ionian Greek from sixth/fifth century B.C. Ephesus. Reformulated to the present context, "a person cannot enter the same city, street, building … twice because…". Different urban elements have different “life expectancy”, that is, time scales: main streets and infrastructures are relatively slow compared, say, to buildings which are slow compared to e.g. flowers in their garden, and so on. But there are also explicit movements, such as traffic, production, in-and outflow of goods, and nowadays most important information production and flows. There are birth-and death processes. Regarding such movements, it is interesting to mention them. On average, however, we may assume that the rates or even rhythms (e.g. day-and night cycle) are relatively time-independent.

The biologist von Bertalanffy (1968) (1901–1972) coined the term “flux equilibrium” to characterize this systemic state. It includes a high coordination of activities, but facilitated by daily routine. The maintenance of the “function” of this “organism” requires a continuous “fight” against the decay of buildings, waste production, etc., in short “entropy production". A fight consuming energy. The existence of the dynamic steady state is possible because of the confidence of the citizens in the stability of the steady state in accordance with our observations (Chap. 7) on SIRNIA and Friston’s Free Energy Principle.While the order parameter is invisible like the “invisible hand” of Adam Smith in economy, we may bring to light the various interdependencies between the relevant indicators.

An example of confidence in a steady state comes from a paper by Ruthen (1993) about the food market of New York City. Here is a quotation about that from Portugali (2000): “a large number of firms of all sizes supply food for over 7 millions people, ‘without creating shortages or surpluses’. All this happens as if by itself, with no central planning nor public authority to regulate the process and safeguard it by producing stocks and keeping of food stores. Furthermore, none of the millions of the interacting elements in the system—individual households, firms of all kinds and restaurants, keep stock for more than a day or two. And yet it works and very efficiently so.”

8.1.4 Urban Rhythms

A nice simple way to study urban rhythms is Hägerstrand’s (1970) *Time Geography*. Mapping the space-time movements performed by humans he suggested a distinction between *daily path* and *life path*. Life paths refer, for instance, to a person who first lives with his/her parents, then as a youngster in a small city center apartment, then with own family in a suburb. As illustrated in Fig. 8.1 *left*, a typical life path is a one directional movement with a few moves, when each such move is a kind of phase transition in the life of the person concerned. For example, living with parents until, say, end of high-school, then a certain period with several moves when living as a single, and living with own family. Daily paths, as the name indicates, are the mapping of the daily movement of a person. Figure 8.1 *right* illustrates a typical daily path of a working person—waking up in the morning, commuting to work, working for say 8 h and then commuting back home and so day after day. As can be seen, the daily movements are essentially repetitive space–time routines.

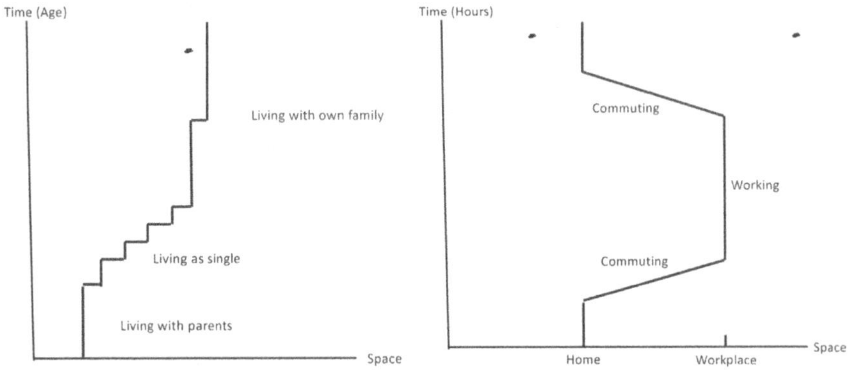

Fig. 8.1 *Left*. A typical life path. *Right*. A typical daily path of a person

Hägerstrand's focus of interest was on the daily paths and routines, suggesting that routinized, daily, space-time movements, are determined by three sets of constraints: *capability* constraints, resulting from humans' basic needs to regularly sleep, eat, etc., as well as from their bodily and technical movement capabilities (walking, biking, driving); *coupling* constraints reflect the socio-economic reality that humans' action (e.g. work) requires collaboration with others in specific places (e.g., workplaces); *authority* constraints, refer to formal laws such as the opening and closing time of shops, but also to cultural habits such as the "siesta" in Spain.

Routinized action, including movement in space, is thus a basic property of human behavior. A real-world event that demonstrates this is the tragic 2010 earthquake in Port-au-Prince, Hawaii. A study by Lu et al. (2012) investigated the population movement following the earthquake by means of big data (cell phones). They have found that contrary to the widely held view that such a large-scale extreme event (e.g. phase transition) must cause people to behave and move chaotically, Lu et al. showed that people's instinctive behavior was to immediately develop routinized space-time behavior that closely followed their past routinized patterns of movement strongly influenced by their historic behavior and their social bonds. In a follow up study, Kennett and Portugali (2012) have further interpreted Lu's et al. study by reference to Hägerstrand's space–time model.

Hägerstrand and time geography studies have treated daily and life paths as independent phenomena. However, as suggested by Portugali (2020/1) and as already hinted at above, in the context of complexity, the two are in fact aspects of a typically behaving complex system with long periods of steady state interrupted by short events of phase transition: life paths are essentially the phase transitions while daily paths the steady states. Furthermore, in terms of the above noted Weidlich's conceptualization, life paths function as OPs for the fast daily paths. Figure 8.2 attempts to capture these relationships: the upper part illustrates the life path (as in Fig. 8.1 *left*), the middle part illustrates the life path in terms of a complex system (steady state, fluctuations followed by phase transition and steady state), while the bottom part represents the daily paths as routinized movements during steady state.

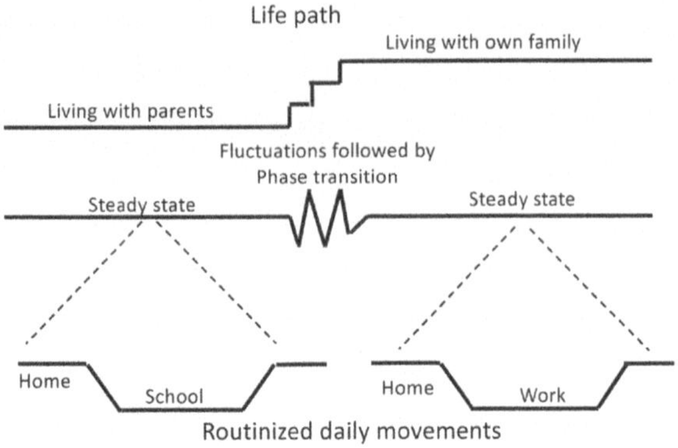

Fig. 8.2 Life and daily paths in terms of complexity/synergetic theories

Life and daily paths focus on single individuals, yet these individual rhythms are the engines of the dynamics of cities. In particular daily paths exhibit high level of a synchronized collective behavior: every morning millions of people wake up more or less at the same time, commute to city centers more or less at the same time, work and then commute back home, once again at more or less at the same time and this is so day after day ….

8.1.5 Links to Friston's FEP, Synergetics' V and SIRNIA

Complex systems have innate tendency to minimize their FE/*V*. The relatively long periods of steady state that characterize the long-term evolution of such systems is the result of this tendency. Steady state is maintained by a routinized space-time behavior and action of the elementary parts of the system concerned. In the case of the socio-spatial entities 'cities', these are the urban agents that by means of their behavior give rise to the circularly causal processes of socio-spatial-material reproduction, that again and again reproduces the city as it was/is. This is achieved by building the material components of cities (buildings, roads, …) in specific ways and locations, by developing formal socio-political institutions (government, judiciary, police,..) that ensure the reproduction of the social structure, as well as by informal ideological-cultural institutions (prevailing beliefs, 'politically correctness'…) that maintains the overall steady state.

In these processes, urban agents' behavior and action is determined, partly by their interaction with other agents, and partly by the SIRNIA processes, that is, by the information the artifact city and the many smaller artifacts of which it is composed, convey to the urban agents.

8.2 Interdependencies in the Steady State

8.2.1 What Kind of Interdependencies?

As noted above, at a first stage, a settlement/city has come into existence by a process presumably similar to a phase transition where one or several OPs emerged that via the slaving principle determine the relations among the agents and also between them and their artifacts (cf. Chaps. 4 and 5). In view of circular causality, the OPs have come into existence just because of these relations, which now form the basis of the evolved steady state and its stability. In this section we deal with the question in how far we can unearth such relations by the analysis of "macroscopic" measurements of indicators/quantifiers. This top–down approach may be an alternative to bottom–up approaches dealing with "microscopic" agent-based models (Chaps. 4 and 5). How can we find a cue to this prospective macroscopic method?

A system in steady state, e.g. a city, is not like a rigid body, but shows a considerable amount of movement, rhythmic and non-rhythmic. But in addition, there are all the time fluctuations, such as birth and death, marriages, heights and lows of sales of all sorts of goods and production processes, heating and cooling of buildings, car accidents etc. All that can be captured by measuring the amount of fluctuations around mean values of the corresponding indicators/ quantifiers. It is the "fluctuation" dynamics inherent in all these processes that helps us to unearth interdependencies.

A first hint at how we may proceed is provided by our above citizen (s)/income (q) model of Sect. 7.3.1. There we derived a joint probability distribution based on the variance of s and q, but, most importantly, on the correlation function between s and q. As our formalism revealed, the mean values of s and q vanish unless there is a "source" term s_0 in the joint probability distribution. So the question arises: is there a general approach by which we can derive the size of this source from measured data, and under which conditions is a second one needed. The detailed answer will be given below.

The approach presented there can, of course, be applied to any pairs of quantifiers. Our goal is, however, more ambitious. Namely, we want to unearth chains of statistical interdependencies. A few rather obvious examples may suffice.

(1) citizens, income, city tax, total income of city,
(2) citizens, vehicles, accidents,
(3) to city imported raw material, produced cars, cars sold to citizens.

Our general approach ("formalism"), will be presented in the following section. To elucidate our approach, we treat scenarios dealing with three quantifiers:

(1) an open-ended chain of quantifiers,
(2) a closed chain (loop), but all "links" (i.e. correlation functions) equal,
(3) a closed chain, but with two different link sizes.

A few notes may be in order. Since all correlation functions are symmetric, we deal only with bidirectional links. They mirror the statistical interdependencies, but not causal relationships. E.g. in the citizen/income example, by knowing the "link" parameter β, we can determine q once we know s, but equally well s, once we know q. (More precisely speaking, we determine, e.g. the probability to find q, once we have measured s, or in still other words, we determine the surprise [cf. Sect. 7.3.1, Eq. (7.19)]. The sources may be part of the system (the city), having come into existence by self-organization, or may be part of the environment, e.g. delivery of goods to the city.

8.2.2 *Formalism*

We treat a system described by J variables, $q_1, q_2, \ldots, q_{\mathcal{J}}$.

We assume that by sampling the following quantities have been measured (where $\langle \ldots \rangle$ denotes averages)

$$\langle q_j \rangle, S_j = \langle q_j^2 \rangle - \langle q_j \rangle^2, \\ C_{j\ell} = \langle q_j q_\ell \rangle - \langle q_j q_\ell \rangle \tag{8.1}$$

The best guess on the underlying probability distribution is

$$p(q) = \exp(-V(q)) \tag{8.2}$$

$$V(q) = \sum_{j=1}^{J} \sum_{\ell=1}^{J} \Lambda_{jl} q_j q_\ell - \sum_{j=1}^{J} \Lambda_j q_j - \Lambda_0 \tag{8.3}$$

Note that $S_j, C_{j\ell}$ are invariant against the substitution

$$q_j \to q_j + b_j; b_j : constant \tag{8.4}$$

$\Lambda = (\Lambda_{j\ell})$ is the inverse of a matrix.

$$M = (M_{j\ell}) \text{ where } M_{jj} = S_j, M_{j\ell} = C_{j\ell}, j \neq \ell \tag{8.5}$$

The eigenvalues λ_k of Λ are just the inverse of eigenvalues μ_k of M, i.e. $\lambda_k = \mu_k^{-1}$. The relation $\Lambda = M^{-1}$ can be proven by mean of the orthonormal transformation

$$q_j = \sum_{k=1}^{J} a_{jk} Q_k \tag{8.6}$$

So that the quadratic/bilinear part of V (8.3) becomes "diagonal".

$$\widetilde{V} = \sum_{k=1}^{J} \lambda_k Q_k^2 - \sum_{j=1}^{J} \Lambda_j \sum_{k=1}^{J} a_{jk} Q_k - \Lambda_0 \tag{8.7}$$

Simultaneously we obtain

$$\langle q_j \rangle = \sum_{j=1}^{J} a_{jk} \langle Q_k \rangle \tag{8.8}$$

And because of

$$\sum_{j=1}^{J} a_{jk} a_{jm} = \delta_{km} \tag{8.9}$$

$$\langle Q_k \rangle = \sum_{j=1}^{J} a_{jk} \langle q_j \rangle \tag{8.10}$$

The mean value $\langle Q_k \rangle$ of a Gaussian with (8.7) is

$$\langle Q_k \rangle = \frac{1}{2\lambda_k} \sum_{j=1}^{J} \Lambda_j a_{jk} \tag{8.11}$$

From (8.10) and (8.11) we obtain

$$\Lambda_j = \sum_{\ell=1}^{J} B_{j\ell} \langle q_\ell \rangle \tag{8.12}$$

where

$$B_{j\ell} = \sum_{k=1}^{J} 2\lambda_k a_{jk} a_{\ell k} \tag{8.13}$$

The relations (8.12) and (8.13) show how Λ_j depends on $\langle q_j \rangle$. Inserting (8.11) in (8.8) leads us to:

$$\langle q_j \rangle = \sum_{\ell=1}^{J} D_{j\ell} \Lambda_\ell \tag{8.14}$$

where

$$D_{j\ell} = \sum_{k=1}^{J} \frac{1}{2\lambda_k} a_{jk} a_{\ell k} \tag{8.15}$$

We call the Λ_ℓ s the sources of q_j. Jointly with the "structural coefficients" a_{jk} the Λ_ℓ s determine how large the $\langle q_j \rangle$ s are. If all sources are zero, so are all mean values. The whole approach requires that $p(q)$ (8.2) can be normalized. This is the case if and only if all eigenvalues of the matrix Λ are positive.

Finally, Λ_0 is fixed by the normalization condition of $p(q)$ or $\tilde{p}\ (Q)$, respectively. The relations (8.6) jointly with the mean and the variance of a_k, $k = 1, -, \mathcal{J}$, determine the hidden inter-dependences among the variables/quantities q_j. In particular, the relations (8.12) and (8.13) allow us to unveil the required sources. In view of practical application, we mention that all mathematical Apps require only linear algebra (can easily implemented on computers). In a number of cases the results can be presented in closed form. We present some explicit examples to elucidate the way we may find the sources.

8.2.3 *Examples*

(1) *Two quantities* q_1, q_2

By a proper scaling of q_1, q_2 we may assume [Eq. (8.1)]

$$\mathcal{S}_1 = 1, \mathcal{S}_2 = 1 \tag{8.16}$$

We abbreviate $C_{12} = C_{21}$ by β.
The matrix M 8.5 reads

$$M = \begin{pmatrix} 1 & \beta \\ \beta & 1 \end{pmatrix} \tag{8.17}$$

eigenvalues are

$$\mu_1 = 1 + \beta, \mu_2 = 1 - \beta, \tag{8.18}$$

the normalized transformation coefficients read

$$a_{11} = \frac{1}{\sqrt{2}} \quad a_{12} = -\frac{1}{\sqrt{2}} \tag{8.19}$$

$$a_{21} = \frac{1}{\sqrt{2}} \quad a_{22} = \frac{1}{\sqrt{2}} \tag{8.20}$$

where the second index refers to the index of μ (8.18).

To unveil the sources according to (8.12) and (8.13), we calculate $B_{j\ell}$. Note that in (8.13) we have to replace λ_k by μ_k^{-1}.

$$\mathcal{B}_{11} = \mathcal{B}_{22} = \frac{2}{1-\beta^2}, \mathcal{B}_{12} = \mathcal{B}_{21} = -\frac{2\beta}{1-\beta^2} \tag{8.21}$$

And obtain

$$\Lambda_1 = \frac{2}{1-\beta^2}(\langle q_1 \rangle - \beta \langle q_2 \rangle) \tag{8.22}$$

$$\Lambda_2 = \frac{2}{1-\beta^2}(-\beta \langle q_1 \rangle + \langle q_2 \rangle) \tag{8.23}$$

By measuring $\langle q_1 \rangle$ and $\langle q_2 \rangle$ and using (8.22) and (8.23), we can determine the sources Λ_1, Λ_2.

In the context of urbanism, we may identify q_2 with city size s and q_1 with some quantities q, e.g. income. Using our example that.

$\langle q \rangle = \beta \langle s \rangle$(or $\langle q_1 \rangle = \beta \langle q_2 \rangle$), we obtain $\Lambda_1 = 0$, so that there is only one source, Λ_2.

Because of its index 2, it belongs to the variable q_2.

If, however, $\langle q_1 \rangle > \beta \langle q_2 \rangle$, then $\Lambda_1 > 0$, so that $\langle q_1 \rangle$ has two sources.

(2) *Three quantities in an open chain*

Like in example (1) we scale the quantities such that

$$S_j = 1, j = 1, 2, 3.$$

In our model we assume that $\mathcal{C}_{12} = \mathcal{C}_{23} = \beta$, but $\mathcal{C}_{13} = 0$. This fixes the elements of $\mathcal{M}$ cf. (8.5). The eigenvalues and transformation elements of $\mathcal{M}$ are

$$\mu_1 = 1 \quad a_{11} = \frac{1}{\sqrt{2}} \quad a_{21} = 0 \quad a_{31} = -\frac{1}{\sqrt{2}} \tag{8.24}$$

$$\mu_2 = 1 + \sqrt{2}\beta \quad a_{12} = \frac{1}{2} \quad a_{22} = \frac{1}{\sqrt{2}} \quad a_{32} = \frac{1}{2} \tag{8.25}$$

$$\mu_3 = 1 - \sqrt{2}\beta \quad a_{13} = -\frac{1}{2} \quad a_{23} = \frac{1}{\sqrt{2}} \quad a_{33} = -\frac{1}{2} \tag{8.26}$$

To derive the coefficients $D_{j\ell}$ in (8.14), i,e.

$$\langle q_j \rangle = \sum_{\ell=1}^{3} D_{j\ell} \Lambda_\ell \tag{8.27}$$

we use (8.15) where we replace $\frac{1}{\lambda_k}$ by μ_k. We obtain

$$D_{j\ell} = \frac{1}{2}\begin{pmatrix} 1 & \beta & 0 \\ \beta & 1 & \beta \\ 0 & \beta & 1 \end{pmatrix} \tag{8.28}$$

To elucidate the result (8.27) and (8.28), we consider two examples where only one source term Λ_j is different from zero.

In the case $\Lambda_1 \neq 0$, $\Lambda_2 = \Lambda_3 = 0$ we obtain:

$$q_1 = \frac{1}{2}\Lambda_1, q_2 = \frac{1}{2}\beta\Lambda_1, q_3 = 0 \tag{8.29}$$

This result is remarkable, because it is counterintuitive. Because of the "causal chain" $\Lambda_1 \to q_1 \to q_2 \to q_3$ with nonvanishing links (correlations $\mathcal{C}_{12}, \mathcal{C}_{23}$) we would have expected that $q_3 \neq 0$ in contrast to (8.29).

What is the reason of this contradiction?. We ignored that the correlations between $1 \to 2$ and $2 \to 3$ also introduce a correlation between 1 and 3 which we should have taken into account by a measured $\mathcal{C}_{13}$. This problem doesn't show up in the case of $\Lambda_2 \neq 0$, $\Lambda_1 = \Lambda_3 = 0$, because there are direct links between q_2 and q_1, q_3.

(3) *Three quantities in a closed chain. Equal links*

We treat the more general case of $\mathcal{L}$ quantities in a cluster, where all quantities are linked with each other. As in the foregoing examples we use normalized quantities so that

$$\mathcal{M}_{jj} = 1 \text{ for } j = 1, \ldots, \mathcal{L} \tag{8.30}$$

Both in the general case and for $\mathcal{L} = 3$, we make the assumption

$$\mathcal{M}_{j\ell} = \beta, \; j \neq \ell \tag{8.31}$$

That is rather stringent, but allows a detailed, explicit treatment. The equations for the transformation coefficients a_{jk} read

$$a_{jk}(1-\beta) + \beta\sum_{m=1}^{\mathcal{L}} a_{mk} = \mu_k a_{jk} \tag{8.32}$$

The eigenvalues are:

$$k = \mathcal{L} \; \mu_{\mathcal{L}} = 1 + \beta(\mathcal{L} - 1) \tag{8.33}$$

$$k \neq \mathcal{L} \; \mu_k = (1 - \beta) \tag{8.34}$$

The normalized coefficients read

$$k = \mathcal{L}\ a_{j\mathcal{L}} = \mathcal{L}^{-\frac{1}{2}} \tag{8.35}$$

$$k \neq \mathcal{L}\ a_{j\mathcal{L}} = \mathcal{L}^{-\frac{1}{2}} \exp\left(\frac{2\pi i k j}{\mathcal{L}}\right) \tag{8.36}$$

By use of (8.15) with $\lambda_k = \mu_k^{-1}$ we obtain

$$\mathcal{D}_{j\ell} = \frac{1}{2}\big((1-\beta)\delta_{j\ell} + \beta\big) \tag{8.37}$$

So that

$$\mathcal{D}_{jj} = \frac{1}{2}, \tag{8.38}$$

$$\mathcal{D}_{j\updownarrow} = \frac{1}{2}\beta,\ j \neq l \tag{8.39}$$

Thus according to (8.6)

$$\langle q_j \rangle = \frac{1}{2}\left(\Lambda_j + \beta \sum_{\ell \neq j} \Lambda_\ell\right) \tag{8.40}$$

Let us consider the special case that only one Λ, say $\Lambda_1 \neq 0$. Then

$$\langle q_1 \rangle = \frac{1}{2}\Lambda_1 \tag{8.41}$$

$$\langle q_j \rangle = 0,\ j = 2, \ldots, \mathcal{L} \tag{8.42}$$

This means that in a cluster, or in a closed chain of three quantities, under the condition

$$\langle q_j \rangle \neq 0 \tag{8.43}$$

each quantity q_j needs its own source to exist.

For sake of completeness, we quote the result for the coefficients $\mathcal{B}_{j\ell}$ that are needed to calculate the Λ s provided the $\langle q_\ell \rangle$ s are known, (8.12) and (8.13).

$$\mathcal{B}_{j\ell} = \frac{2}{1 + \beta(\mathcal{L}-2) - \beta^2(\mathcal{L}-1)\mathcal{L}}\big(\delta_{j\ell}(1 + \beta(\mathcal{L}-1)) - \beta\big) \tag{8.44}$$

(4) *Three quantities in a closed chain. Two different links*

We use normalized quantities so that $\mathcal{M}_{jj} = 1$, $j = 1, 2, 3$. We treat the following correlations $\mathcal{C}_{j\ell} = \mathcal{M}_{j\ell}$.

$$\mathcal{M}_{12} = \mathcal{M}_{21} = \mathcal{M}_{23} = \mathcal{M}_{32} = \beta, 0 < \beta < 1 \tag{8.45}$$

And

$$\mathcal{M}_{13} = \mathcal{M}_{31} = \beta^2 \tag{8.46}$$

The choice (8.46) differs from (8.31). We make it, because it allows us to bring out an essential difference between (3) and 4) (for details cf. below), and it is still simple enough. The equations for the transformation coefficients a_{jk} (or in short a_j) read

$$\mu a_1 = a_1 + \beta a_2 + \beta^2 a_3 \tag{8.47}$$

$$\mu a_2 = \beta a_1 + a_2 + \beta a_3 \tag{8.48}$$

$$\mu a_3 = \beta^2 a_1 + \beta a_2 + a_3 \tag{8.49}$$

The eigenvalues and solutions read, by use of the abbreviation

$$\Gamma = (\beta^2 + 8)^{\frac{1}{2}} \tag{8.50}$$

$$\begin{aligned} &\mu_1 = 1 + \frac{\beta^2}{2} - \frac{1}{2}\beta\Gamma \\ &a_{11} = \mathcal{N}_1 \quad a_{21} = -\frac{1}{2}\mathcal{N}_1(\beta + \Gamma) \quad a_{31} = \mathcal{N}_1 \end{aligned} \tag{8.51}$$

$$\begin{aligned} &\mu_2 = 1 + \frac{\beta^2}{2} + \frac{1}{2}\beta\Gamma \\ &a_{12} = \mathcal{N}_2 \quad a_{22} = -\frac{1}{2}\mathcal{N}_2(\beta - \Gamma) \quad a_{32} = \mathcal{N}_2 \end{aligned} \tag{8.52}$$

$$\begin{aligned} &\mu_3 = 1 - \beta^2 \\ &a_{13} = \mathcal{N}_3 \quad a_{23} = 0 \quad a_{33} = -\mathcal{N}_3 \end{aligned} \tag{8.53}$$

The normalization constants $\mathcal{N}_j$ are fixed by

$$\mathcal{N}_1^{-2} = 2 + \frac{1}{2}(\beta^2 + \beta\Gamma + 4) \tag{8.54}$$

$$\mathcal{N}_2^{-2} = 2 + \frac{1}{2}(\beta^2 - \beta\Gamma + 4) \tag{8.55}$$

$$\mathcal{N}_3^{-2} = 2 \tag{8.56}$$

The coefficients $\mathcal{D}_{j\ell}$ (8.15) can be written in the form

$$D_{j\ell} = G\big(Aa_{j1}a_{\ell 1} + Ba_{j2}a_{\ell 2} + Ca_{j3}a_{\ell 2}\big) \tag{8.57}$$

where

$$G = \frac{1}{2}\mathcal{N}_1^2\mathcal{N}_2^2\mathcal{N}_3^2 > 0 \tag{8.58}$$

and

$$\mathcal{A} = 2\mu_1\mathcal{N}_2^{-2} = 2\left(a - \frac{1}{2}\beta\Gamma\right)\left(b - \frac{1}{2}\beta\Gamma\right) \tag{8.59}$$

where

$$a = 1 + \frac{\beta^2}{2}, b = 4 + \frac{1}{2}\beta^2, \Gamma \text{cf.}(8.50) \tag{8.60}$$

and

$$\mathcal{B}(\Gamma) = \mathcal{A}(-\Gamma) \tag{8.61}$$

The coefficients $\mathcal{D}_{j\ell}$ can be calculated in a straight forward manner by inserting (8.51)–(8.56) and (8.58–8.61) into (8.57) and rearranging terms. We quote only results that allow us to bring out the essentials.

From (8.57) there follows directly.

$$\mathcal{D}_{j\ell} = \mathcal{D}_{\ell j} \text{ for all } l \text{ and } j \tag{8.62}$$

Furthermore we find

$$\mathcal{D}_{33} = \mathcal{D}_{11}, \mathcal{D}_{23} = \mathcal{D}_{12} \tag{8.63}$$

Because the explicit expression for $\mathcal{D}_{j\ell}$ are somewhat lengthy, we present only the leading terms according to the powers of β $(0 < \beta < 1)$.

$$\mathcal{D}_{11} = G32 \tag{8.64}$$

$$\mathcal{D}_{22} = G32 \tag{8.65}$$

$$\mathcal{D}_{33} = G32 \tag{8.66}$$

$$\mathcal{D}_{12} = -G8\beta \tag{8.67}$$

$$\mathcal{D}_{13} = G32\beta^2 \tag{8.68}$$

$$\mathcal{D}_{12} = -G8\beta \tag{8.69}$$

The exact factors of G contain terms of order β^2 and higher.

Now we have all ingredients together to draw our conclusion. We consider (8.14) that connects the sources Λ_ℓ with the mean values $\langle q_j \rangle$. We are particularly interested whether there are chains between the mean values based on a single source Λ, say Λ_1. In this case, by use of our above results, the relations (8.14) read

$$\langle q_1 \rangle = G32\Lambda_1, \langle q_2 \rangle = -G8\beta\Lambda_1, \langle q_3 \rangle = G32\beta^2\Lambda_1 \tag{8.70}$$

The factors $32G$, $8G$ are of no relevance for our discussion. Rather we note that there exists a chain

$$\Lambda_1 \to \langle q_1 \rangle \to \langle q_2 \rangle \to \langle q_3 \rangle \tag{8.71}$$

(Note that our approach allows negative values of the mean values). If in measurements all $\langle q_j \rangle > 0$, then we must dismiss this chain and include more Λ s.

8.3 Conclusions

Similarly to complex systems in general, cities are characterized by long periods of StS interrupted by short events of PT. As noted at the outset of this chapter, despite of this, and the centrality of StS in the dynamics and life of cities, the topic has received only marginal attention. This chapter is a first attempt to correct this situation by elucidating the reasons (or "mechanism") for the long periods of StS that characterizes the evolution of cities. It must be added that in the context of cities, in a number of cases, the notion StS must be taken with a grain of salt, because the population size may grow (or shrink). The point is that the change is so slow that the interdependencies discussed in this chapter are conserved. This is analogous to the "adaptive approximation physics".

In the first part of the chapter, our basic answer rested on the general Synergetics' relationships elaborated in Chap. 3: OP → slaving → circular causality → StS. We've first explored the role of urban rhythms in preserving StS. Then, we've added insight from the discussion in Chap. 7 regarding Friston's FEP and the Synergetic potential V: Both indicate that complex systems have an innate tendency to minimize their "free energy" and the potential V, implying a tendency to avoid or minimize surprise and to maintain StS.

In the second part of the paper, we have tried to shed light on a more detailed, explicit mechanism, namely the numerous specific interdependencies among the activities of the urban agents as they take place in the city. In order to unearth them we have presented a formalism that in the spirit pf Synergetics' 2nd Foundation proceeds as follows: based on statistical data on the strength of the considered activities and of pairwise correlations, we make a best guess (sensu Jaynes) on the overall probability distribution of activities. From it, we then derive interdependency chains, which shed light on these interrelations.

While our formalism can easily be implemented on computers, we should discuss its limitations and how they might be overcome by future research.

(1) We take only pairwise correlations into account.
(2) The probability distribution $p(q)$ represents statistical, but not causal relations.
(3) We discuss interdependencies but not their connection to cities' space-time rhythms.

Regarding limitation (1), one may argue that higher order moments and correlations are rare and can be neglected. This is (most probably) so in the case of a stable steady state, whereas in case of the occurrence of a PT higher order terms are needed (cf. Chaps. 9 and 10).

Regarding limitation (2), to reveal causal relations, time-dependent correlation functions are needed. On their basis, we may derive Fokker-Planck equation (cf. Chap. 5). If it is linear, it represents an Orenstein-Uhlenbeck process and can be treated analytically, otherwise numerical methods must be invoked.

Regarding limitation (3), to explore the links between activities' interdependencies and urban rhythms, we'll need to reveal, time dependent (as we did above) and subsequently causal relations as just noted with respect to limitation (2). As just mentioned, these extensions will have to await further research.

References

Batty, M. (2013). *The new science of cities*. Cambridge Mass: MIT Press.
Bertalanffy von, L. (1968).*General system theory*. New York: George Braziller.
Bettencourt, L. M. A. (2013). The origins of scaling in cities. *Science, 340,* 1438–1441.
de Certeau, M. (1984). *The practice of everyday life*. Berkeley: University of California Press.
Eldredge, N., & Gould, S. J. (1972). Punctuated equilibria: An alternative to phyletic gradualism. In T. J. M. Schopf (Ed.), *Models in Paleobiology* (pp. 82–115). San Francisco: Freeman Cooper.

Giddens, A. (1979). *Central problems in social theory: Action, structure and contradiction in social analysis*. London: Macmillan.

Giddens, A. (1984). *The constitution of society. Outline of the theory of structuration*. Cambridge:Polity.

Goffman, E. (1956). *The presentation of self in everyday life*. US: Doubleday.

Gould, S. J. (1991). "Opus 200," Natural history, p. 16.

Hägerstrand, T. (1970). What about people in regional science? *Papers in Regional Science, 24*, 7–21.

Kennett, D., & Portugali, J. (2012). Population movement under extreme events. *PNAS, 109*, 11472–11473.

Lefebvre, H. (1968/1994). *Everyday life in modern world*. In S. Rabinovitch (Ed.), Translator. Transactions. New Brunswick. First published in 1968 by Editions Gallimard, Paris, as *La vie quotidienne dans le monde moderne*.

Lu, X., Bengtsson, L., & Holme, P. (2012). Predictability of population displacement after the 2010 Haiti earthquake. *Proceedings of National Academy Science,109*, 11576–11581.

Portugali, J. (2000). *Self-organization and the city*. Berlin/Heidelberg/New York: Springer.

Portugali, J. (2011). *Complexity, cognition and the city*. Berlin/Heidelberg/New York: Springer.

Portugali, J. (2020/1). *The second urban revolution* (in Preparation).

Ruthen, R. (1993). Adapting to complexity. *Scientific American, 268*, 11Q – 117.

Chapter 9
Phase Transitions

9.1 Introduction

A central methodological principle of Synergetics is to "look for qualitative changes at macroscopic scales" (Haken 1996, p. 39). When we do so we realize that complex systems are characterized by relatively long periods of steady state (StS) interrupted by short events of phase transitions. StS and PTs are thus among the central concepts of complex systems. And yet, as noted in the previous Chap. 8 (Sect. 8.1.1), "looking at the evolving discourse in the domain of CTC, it can be observed that compared to the other CTC notions, StS and PT have received only marginal attention… [and] when StS and PT are being referred to in CTC, they are presented in the most general way …". As further emphasized there, there is thus a need to elaborate on the role of StS and PT in the dynamics of cities. In Chap. 8 our focus was on the dynamics of StS. In the present discussion we elaborate on the *general* properties and dynamics of PTs, while in Chap. 10 that follows, on PTs in cities.

9.2 Phase Transitions Are Ubiquitous

Phase transition (PT): the process by which a system undergoes a qualitative change. Such changes are observed in quite different fields.

Physics: We consider matter that is composed of very many atoms or molecules such as H_2O. They many appear in different phases of "aggregation": Solid (ice), fluid (water), gaseous (water vapor). Though the individual constituents are always the same, e.g. H_2O, the macroscopic properties of the phases differ qualitatively. Obviously the mechanical properties are quite different, but many others also, such as thermal conductivity, specific heat and so on.

But not only the macroscopic features differ, but also the underlying microscopic structures, i.e. the spatial arrangement of atoms/molecules and their mutual motions.

H. Haken and J. Portugali, *Synergetic Cities: Information, Steady State and Phase Transition*, Springer Series in Synergetics,
https://doi.org/10.1007/978-3-030-63457-5_9

Ice is a highly regular crystal that is quite rigid, the molecules can perform only small oscillations around their equilibrium positions. In water, the molecules may move around but maintain some average distance so that a fixed density results. In the vapor phase (gas) the molecules freely fly around, collide with each other whereby they exchange energy and momentum.

Quite remarkably, transitions between phases (PT) occur at a "critical temperature T_c" (other parameters being fixed). Thus we speak of freezing temperature, boiling temperature.

Some further examples of PT in physics are:

Ferromagnetism: a ferromagnet such as the magnetic needle of a compass possesses a magnetic field at room temperature. This field disappears when the magnet is heated above a critical temperature T_c (Curie temperature). At the microscopic level, the ferromagnet is visualized as a crystal—like arrangement of tiny elementary magnets. Each of them can be represented by an arrow with the "north pole" at its top. In the magnetic state, all these arrows (or most of them) point in the same direction, whereas in the state above T_c they randomly point into all directions (or, if only two directions, up and down, are permitted, randomly up and down).

Superconductivity: Another well-known example is provided by superconductivity. Some electric conductors lose their resistance completely below a critical temperature. E.g. in a ring, the current will flow forever. In a large class of superconductors, at the microscopic level the infinitely high conductivity is attributed to the formation of pairs of electrons leading to a new form of matter. In other classes of superconductors, the processes at the microlevel are not yet fully understood. All the above mentioned PTs occur in closed systems in thermal equilibrium. There is another class of PTs that occur in open system. These systems are driven by an influx of energy, matter and/or information and corresponding outflux (in modern term "waste disposal").

A prominent example of a nonequilibrium phase transition (NPT) is provided by the transition of the disordered light of a conventional lamp to the highly ordered, coherent laser light. We have discussed this phenomenon in Chap. 3. Fluid dynamics offers us further examples of NTPs, such as the Bénard instability. Here, in a circular or rectangular vessel a fluid layer is heated from below and cooled from above so that a temperature difference ΔT_c ("gradient") between the lower and upper surface is generated that serves as control parameter (cf. Sect. 5.1 of Chap. 5). Below a critical temperature difference ΔT_c, heat is transported from below to above by microscopic processes, but above ΔT_c by a macroscopic motion (convection). At the macroscopic level, visible macroscopic patterns of fluid motions appear, e.g. hexagons, stripes, or squares.

A typical feature of NPTs is the existence of PT-hierarchies: with increase of a typical control parameter C, at a sequence of specific values of C, new spatio-temporal patterns appear. E.g. in the laser at some pump power (energy input) laser action sets in, then at an elevated level oscillations may occur, and eventually "deterministic chaos". Similarly, in a typical "Bénard"—experiment, first a roll pattern appears, then at an increased ΔT value an oscillation of roles, then more involved oscillations of roles, and eventually deterministic chaos (a weak form of turbulence). At least the first NPTs in different systems (lasers, fluids) share an important feature: the efficiency increases suddenly (laser: output/input power, fluid: heat transport).

Biology abounds of phenomena that we may conceive as NPTs, though they go on much longer time scales than those in physics. Let us just mention metamorphosis: egg, caterpillar, pupa, insect. These stages largely differ from each other both microscopically and macroscopically. An intriguing question coming from urbanism is: how does the organism manage to stay alive in spite of the dramatic changes and how does it manage them by self-organization?

At least a number of processes in cognition may be understood as NPTs.

A striking example is the recognition of an Dalmatian dog (Fig. 9.1). Here, out of seemingly irregular dots, suddenly the percept "dog" emerges (Gregory 1970). As Dehaene (2014) and coworkers have shown by means of sophisticated experiments, in the human brain, the becoming conscious of some optical signal, e.g. an image, strongly resembles a NPT. Later on, we will deal with NTPs in cities. Before that, to prepare the ground, we discuss characteristic features of both PTs and NPTs. Then we will provide the reader with an outline of the underlying theories.

Characteristic features of PTs and NPTs are: when a system that is close to a critical point (e.g. T_c) is disturbed, it relaxes more slowly to its equilibrium state than when being farther assay. This phenomenon is called "critical slowing down". Furthermore the typical variables of a system close to its critical point show enhanced fluctuations called "critical fluctuations". Both critical fluctuations and critical slowing down may be used as indicators that a system is at or approaching its critical point. Just to mention an example from medicine: Critical fluctuations of EEG activity of a human brain may predict an epileptic seizure. It is speculated that critical fluctuations of seismic activity may be precursors of earthquakes which is of relevance for urbanism.

An important class of PTs and NPTs exhibits a further important feature: "symmetry breaking". Some hints may suffice here. In a ferromagnet, the elementary magnet may point up *or* down ("symmetry"), but in reality they can collectively

Fig. 9.1 The dalmatian dog illusion

realize *only one* direction, e.g. "up". ("Symmetry breaking"). In a fluid cell heated from below, a roll can move clockwise or anticlockwise—but only one direction is realized. Compare also our swimmer example of Chap. 3. In the laser light wave, the phase is arbitrary.

9.3 Some Basic Concepts of PT Theories

As we have seen in Sect. 9.2, at the phase transition point, several properties of a system change dramatically. A prototypical example in the change of magnetization $\mathcal{M}$ of a ferromagnet, where $\mathcal{M}$ become zero above a critical temperature $\mathcal{T}_c$. In the last century, the way $\mathcal{M}$ approaches zero when $\mathcal{T}$ approaches $\mathcal{T}_c$ from $\mathcal{T} < \mathcal{T}_c$ was subject of precise measurements and profound theories (cf. Fig. 9.2). The basic problems are as follows. How can we represent the graph $\mathcal{M}(\mathcal{T})/\mathcal{M}(0)$ versus $\mathcal{T}/\mathcal{T}_c$ by a simple formula in the neighborhood of $\mathcal{T}/\mathcal{T}_c = 1$?

And, can we derive such as a formula and similar ones for other properties by a theory? While the solution to the first task is rather simple (see below), the second task turned out to be highly demanding so that we can present only a brief outline (Figs. 9.2 and 9.3).

To deal with the first task, consider Fig. 9.3 *left*. It is the same as Fig. 9.2 but with different notations $x = \mathcal{T}/\mathcal{T}_c$, $y = \mathcal{M}(\mathcal{T})/\mathcal{M}(0)$ and the corners designated by a, b, c, d. Figure 9.3 *right*, shows Fig. 9.2 but rotated by 90°.

In the neighborhood of $x = 1$ and correspondingly $y = 0$, the graph is reminding us of a parabola and can be approximated by

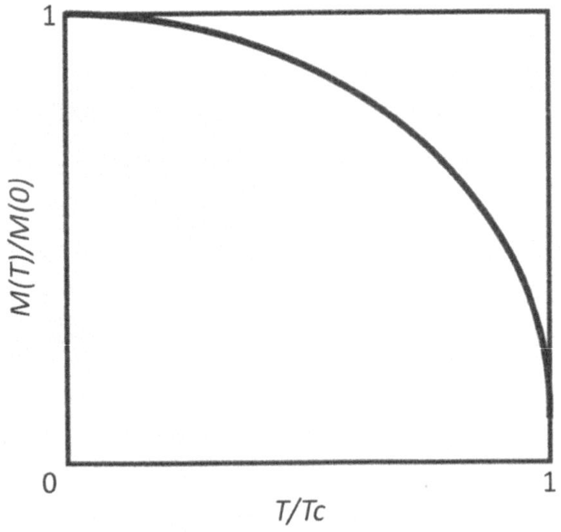

Fig. 9.2 Plot (graph) of $\mathcal{M}(\mathcal{T})/\mathcal{M}(0)$ versus $\mathcal{T}/\mathcal{T}_c$. This diagram is only qualitative but quantitative plots are available

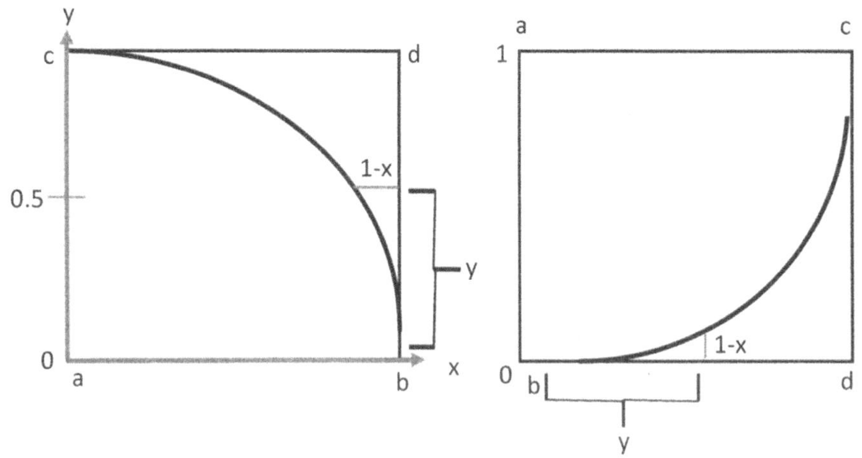

Fig. 9.3 *Left*: same as Fig. 9.2 but with different notations (cf. text). *Right:* Fig. 9.2, but rotated by 90°

$$(1 - x) = y^{\alpha} \tag{9.1}$$

where the exponent is a positive number. Inserting $x = \mathcal{T}/\mathcal{T}_c$, $y = \mathcal{M}(\mathcal{T})/\mathcal{M}(0)$ in (9.1) and resolving for y results in

$$\frac{\mathcal{M}(\mathcal{T})}{\mathcal{M}(0)} = \left(1 - \frac{\mathcal{T}}{\mathcal{T}_c}\right)^{\beta}, \quad \text{where } \beta = \frac{1}{\alpha} \tag{9.2}$$

β is called a *critical exponent*. It has been measured very carefully. By a slight rearrangement of terms, (9.2) can be written as:

$$\mathcal{M}(T) = Const.(\mathcal{T}_c - \mathcal{T})^{\beta} \tag{9.3}$$

This is called "scaling law": The magnetization $\mathcal{M}$ "scales" with the temperature difference $\mathcal{T}_c - \mathcal{T}$ by a "power law".

How can this experimentally established law be derived by a theory?

A famous attempt was made by Landau (1908–1968) with his phase transition theory (Landau and Lifshitz 1980). According to thermodynamics (the science of heat) and statistical physics (physics based on probability theory), the probability p that a physical system is in a specific state is given by the formula

$$p(state) = \mathcal{N}\exp\left(-\frac{\mathcal{F}(state)}{k\mathcal{T}}\right) \tag{9.4}$$

where $\mathcal{N}$ is a (normalization) constant, $\mathcal{F}$ the "free energy" (a thermodynamic concept), k the "Boltzman constant", and $\mathcal{T}$ the absolute temperature.

Landau assumed that the "state" is characterized by the magnetization $\mathcal{M}$ so that (9.4) must read

$$p(\mathcal{M}) = \mathcal{N}exp\left(-\frac{\mathcal{F}(\mathcal{M})}{kT}\right) \tag{9.5}$$

Since in a ferromagnet the magnetizations $+\mathcal{M}$ and $-\mathcal{M}$ are equally probable, p and also $\mathcal{F}$ can depend only on even powers, $\mathcal{M}^2$, $\mathcal{M}^4$,.... Following Landau we put

$$\mathcal{F} = \alpha(\mathcal{T} - \mathcal{T}_c)\mathcal{M}^2 + \beta\mathcal{M}^4 \tag{9.6}$$

where $\alpha > 0$, $\beta > 0$ are parameters. Most importantly, the prefactor of $\mathcal{M}^2$ changes its sign at $\mathcal{T} = \mathcal{T}_c$

$$\alpha(\mathcal{T} - \mathcal{T}_c) < 0 \quad \text{for} \quad \mathcal{T} < \mathcal{T}_c \tag{9.7}$$

$$\alpha(\mathcal{T} - \mathcal{T}_c) > 0 \quad \text{for} \quad \mathcal{T} > \mathcal{T}_c \tag{9.8}$$

$\mathcal{F}$ corresponds to $\mathcal{V}$ of Fig. 5.1. The most probable state $\mathcal{M}$ is where p has its maximum and $\mathcal{F}$ its minimum.

$$\frac{d\mathcal{F}}{d\mathcal{M}} = 0 \tag{9.9}$$

For $\mathcal{T} < \mathcal{T}_c$ we obtain

$$\mathcal{M} = \left(\frac{\alpha}{2\beta}\right)^{\frac{1}{2}} (\mathcal{T}_c - \mathcal{T})^{\frac{1}{2}} \tag{9.10}$$

(where we have chosen the example $\mathcal{M} > 0$).

A comparison between Landau's result (9.10) and (9.2) leads to the prediction of the critical exponent

$$\beta = \frac{1}{2} \tag{9.11}$$

Which does not agree with the experimental value. So the development of the correct theory became a real challenge. It required sophisticated methods of theoretical physics.

Interlude: fractals and self-similarity

Fractals were first conceived by Cantor (1845–1918), while the denotation "fractal" is due to Mandelbrot (1983). A prototypical example of a fractal is the *Cantor set* that can be constructed as follows: Consider the individual 2-D top layer segment

Fig. 9.4 The construction of a cantor set

of Fig. 9.4. Then we take one third, middle part out so that the 2nd layer results. The 3rd 4th 5th and 6th layers show what happens when we continue this procedure. The wanted *fractal* results when we continue this procedure ad infinitum. When we watch this process with our eyes, because of their limited power of resolution there will be an iteration step where we can no more recognize the finer structure.

Just for illustration: We may still recognize the structure of the 6th layer or maybe also the 7th, but no more still finer structures. But we may use a magnifying glass that amplifies the 6th layer so that the four bars on the left side just coincide with the 8 areas of the previous layer. The patterns as can be seen are self-similar. Note that we are interested only in the *line squares* on the x-axis "one-dimensional case". But in precisely the same way we can construct fractals in two or three dimensions. Apropos dimension: the fractals are embedded in the corresponding one-,two- or three- dimensional spaces (line, square, cube). The dimension of fractal on the line is <1, it's fractal dimension. The fractal structure of cities has been considered e.g. by Frankhauser (1991) with respect to the network of streets and by Batty and Longly (1994). Bettencourt et al. (2007), West and Brown (1997) and others have shown that the blood vessels of human and animals possess a fractal structure. Our little "interlude" will help us to explain important concepts of the microscopic theory of the ferromagnetic phase transition that we discuss next.

9.4 Microscopic Theory of the Ferromagnetic Phase Transition

Following the general parlance of this field we speak of "spins" instead of "elementary magnets". These "objects" share the same physical properties needed for our theory. They differ only by some factor. Both objects can be represented by little arrows that may point up or down. The microscopic theory treats all the configurations of spins of which some number $\mathcal{N}_u$ shows *up* and another one, $\mathcal{N}_d$, *down*. The total number of spins $\mathcal{N}$ is $\mathcal{N} = \mathcal{N}_u + \mathcal{N}_d$. The maximal magnetization is obtained when all spins (arrows) point in the same direction. In the *Ising* model of ferromagnetism is assumed that *neighboring* spins interact with each other: their common energy is negative, $-J$, if they point in the *same* direction and positive, $+J$, if they point in *opposite* direction.

Since physical systems in equilibrium, at absolute temperature $T = 0$, realize the state with lowest energy, all spins point in the same direction, so that the magnetization is maximal (Fig. 9.5). When we bring the ferromagnet in contact with a surrounding at a temperature $T > 0$, the ferromagnet is heated up. But according to thermodynamics, "heat" means microscopic irregular (random) motion of atoms or, in the present case, of spins. This means that the spins can now be flipped say "up" to "down" which requires energy (Fig. 9.6). Thus the spins are subject to two forces:

(1) the order generating attractive force.
(2) the order destroying flipping—i.e. the thermal agitation.

A glimpse at Fig. 9.2 gives us a first idea what happens when we increase the temperature. We start from the upper left corner with maximal magnetization $\mathcal{M}$.

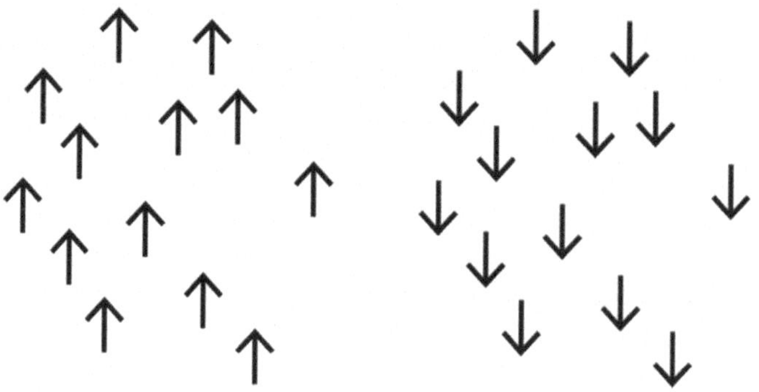

Fig. 9.5 Two states of spins arranged in ferromagnet. At absolute temperature T = 0 all spions point in the *same* direction, up or down

Fig. 9.6 A random configuration at elevated temperature

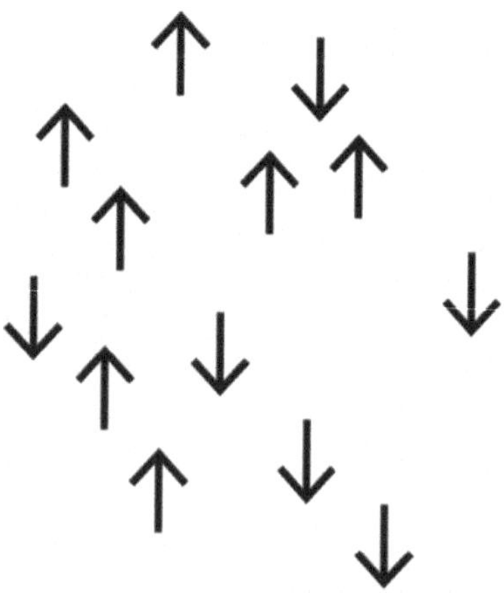

With increasing temperature, $\mathcal{M}$ decreases but little. This means that comparatively only few spins are flipped—until we reach a region close to $\mathcal{T} = \mathcal{T}_c$ where $\mathcal{M}$ decreases strongly so that a high percentage of spins flip—until at $\mathcal{T} = \mathcal{T}_c$ with $\mathcal{M} = 0$ the numbers of "up" and "down" spins become equal. This equal number of spins "up" and "down" can be realized by a variety of spatial configurations. To visualize their nature we consider the two-dimensional *Ising* model: a sheet (or layer) of spins in two dimensions. When we mark on a picture of a spin-arrangement each up spin by a black spot, we will see a mosaic of black and white regions. Now the crucial point in this: Computer Simulations reveal that close to $\mathcal{T} = \mathcal{T}_c$ the spin-configurations strongly resemble a fractal. I.e. if we magnify a part of the picture, a similar picture results ("*Self-similarity*"). This insight provided a key to an adequate treatment of the ferromagnetic phase transition. Leo Kadanoff (1960) "inverted" so to speak the construction of a fractal, that we have described above in our "interlude" by consecutively forming blocks of spins. He starts from a grid where each cell is occupied by one spin. The final derivation of the scaling law (3) with the correct β was achieved by Kenneth Wilson (1983) with his "renormalizations group". Now back to Kadadanoff's approach. Two neighboring spins have an interaction energy $\mp \mathcal{J}$ depending on their relative orientation "parallel" or "antiparallel". $\mathcal{J}$ is the interaction constant. The total energy of each spin configuration depends on this $\mathcal{J}$. The free energy $\mathcal{F}$ can be interpreted as a temperature dependent average over the energies of all (realized) configurations so that $\mathcal{F}$ depends on $\mathcal{J}$ and T, $\mathcal{F} = \mathcal{F}(T, \mathcal{J})$. Now we consider a block of 3×3 spins (in two dimensions) and replace it by a single spin that points in the same direction as the majority of spins of this block. Simultaneously the original grid in shrunk by a factor of $\frac{1}{3}$ so that we again arrive at a spin-lattice of the former size. We can continue this procedure by again and again forming blocks etc. So what is the problem? We require that at each step the same numerical value of $\mathcal{F}$ results.

As a detailed analysis shows, this requires that we replace the initial interaction constant $\mathcal{J}$ between only two neighboring spins by a whole set of interaction constants $\mathcal{K}$ between not only neighbors, but also between more distant pairs of the artificially "created" (defined) spins. Furthermore, how do the constants $\mathcal{K}$ of iteration step $\ell + 1$, i.e. $\mathcal{K}(\ell + 1)$ depend on $\mathcal{K}(\ell)$ of ℓ? Or written as "Symbolic" formula

$$\mathcal{K}(\ell + 1) = R(\mathcal{K}(\ell)) \tag{9.12}$$

("Symbolic" because we are dealing with a whole set of $\mathcal{K}$ s). The above questions posed a highly demanding mathematical problem. But to make a long story short, taken at face value the insights gained are: [we essentially follow the approach by Cardy (1996)].

(1) The step-wise iteration comes to an end, when $\mathcal{K}$ reaches a "fixed point" $\mathcal{K}^*$ where

$$\mathcal{K}^* = \mathcal{R}(\mathcal{K}^*) \tag{9.13}$$

(2) Close to $\mathcal{K}^*$, only two constants $\mathcal{K}$ are important. They can be chosen such that one remains the same when all spins are flipped, we call it $\mathcal{K}_t$. The other one changes its sign; we call it $\mathcal{K}_h$. It contributes to the free energy, if the ferromagnet is subjected to an external magnetic field. The formely "symbolic" equation reads explicitly

$$\mathcal{K}_t(\ell + 1) = \mathcal{R}_t(\mathcal{K}_t(\ell)) \tag{9.14}$$

$$\mathcal{K}_h(\ell + 1) = \mathcal{R}_h(\mathcal{K}_h(\ell)) \tag{9.15}$$

(3) Close to $\mathcal{K}^*$, the equation can be linearized around $\mathcal{K}^*$. E.g.

$$\mathcal{K}_t(\ell + 1) = \mathcal{R}_t(\mathcal{K}_t^*) + \lambda_t(\mathcal{K}_t(\ell) - \mathcal{K}_t^*) \tag{9.16}$$

Because $\mathcal{K}^*$ is a fixed point, (9.16) can be rearranged to

$$\mathcal{K}_t(\ell + 1) - \mathcal{K}_t^* = \lambda_t(\mathcal{K}_t(\ell) - \mathcal{K}_t^*) \tag{9.17}$$

We denote the distance of $\mathcal{K}_t(\ell)$ from the fixed point by $u_t(\ell)$

$$u_t(\ell) = \mathcal{K}_t(\ell) - \mathcal{K}_t^* \tag{9.18}$$

So that (9.17) becomes

$$u_t(\ell + 1) = \lambda_t u_t(\ell) \tag{9.19}$$

In the literature, λ_t is written as b^{y_t} where b is the block size and y_t is called "renormalization group eigenvalue".

Equation (9.19) can be n times iterated so that

$$u_t(\ell + n) = b^{n y_t} u_t(\ell) \tag{9.20}$$

For what follows the explicit value of ℓ is irrelevant; all what counts is our assumption that u_t is small. Therefore we introduce a short hand notation by just dropping "ℓ" so that (9.21) becomes

$$u_t(n) = b^{n y_t} u_t \tag{9.21}$$

(Strictly we put $u_t(\ell + n) = \widehat{u_t}$ and drop ^)

As stated above $\mathcal{K}_t$ and thus $\mathcal{M}_t$ remain invariant against spin-flips. Thus u_t cannot depend on h in leading approximation, but must only depend on temperature. Since u_t vanishes at the fixed point (that we identify with the critical point), we put

$$u_t = \mathcal{C}_1(\mathcal{T}_C - \mathcal{T}) \tag{9.22}$$

where $\mathcal{C}_1$ is a constant.

By the same procedure we obtain

$$u_h(n) = \mathcal{C}_2 h b^{n y_h} \equiv u_h b^{n y_h} \tag{9.23}$$

(4) After these preparations a crucial question arises: The final result, e.g. the free energy, must not depend on the "iteration" parameter *n*.

How can we eliminate it from (9.21) and (9.23)? To this end we choose a fixed, i.e. $\mathcal{T}$—independent constant $\mathcal{C}$ and require

$$(9.21) = \mathcal{C} \tag{9.24}$$

Because $u_t(n)$ must be small enough so that $\mathcal{R}_t$ could be linearized, $\mathcal{C}$ must also be small. The solution b^n to (9.24) with (9.22) reads

$$b^n = \left(\frac{u_t}{\mathcal{C}}\right)^{-\frac{1}{y_t}} \tag{9.25}$$

This means, that n is $\mathcal{T}$-dependent, a statement that usually is "Wiped under the carpet". Inserting (9.25) in (9.23) yields

$$u_h = \mathcal{C}_2 h \left(\frac{u_t}{\mathcal{C}}\right)^{-\frac{y_h}{y_t}} \tag{9.26}$$

(5) To finish our analysis, we consider the behavior of the free energy under "block-spin" transformation. We define the reduced free energy per site, p, by

$$f = \mathcal{N}^{-1} F \tag{9.27}$$

$\mathcal{N}$ total number sites. After the ℓ th iteration, $\mathcal{F}$ and f depend on a set of coupling constants $\mathcal{K}$

$$f(\mathcal{K}) \equiv f(u_t, u_h)$$

When we iterate one step further, then because of (9.21) and (9.23)

$$f(u_h, u_t) = b^{-d} f\left(b^{y_t} u_t, b^{y_h} u_h\right) \tag{9.28}$$

where b is the block size and d the dimension of the ferromagnet.

After n iterations (9.28) becomes

$$f(u_t, u_h) = \underbrace{b^{-nd}}_{\text{I}} f\left(\underbrace{b^{ny_t} u_t}_{\text{II}}, \underbrace{b^{ny_h} u_h}_{\text{III}}\right) \tag{9.29}$$

By use of (9.25) we eliminate b^n which yields

$$\text{I} = \left(\frac{u_t}{\mathcal{C}}\right)^{\frac{d}{y_t}} \tag{9.30}$$

$$\text{II} = \mathcal{C} \tag{9.31}$$

$$\text{III} = \mathcal{C}_2 u_h \left(\frac{u_t}{\mathcal{C}}\right)^{-\frac{y_h}{y_t}} \tag{9.32}$$

Now we are quite close to our goal: the functional dependence of the free energy on temperature and magnetic field. We just have to insert the expressions for u_t and u_h, i.e. (9.22) and (9.23) in (9.30)–(9.32). By a change of the denotations of the constants we obtain

$$f = \left(\frac{\mathcal{T}_c - \mathcal{T}}{t_0}\right)^{\frac{d}{y_t}} \tilde{f}\left(\frac{h}{h_0}\left(\frac{\mathcal{T}_c - \mathcal{T}}{t_0}\right)^{-\frac{y_h}{y_t}}\right) \tag{9.33}$$

This result may be used to derive scaling laws close to $\mathcal{T} = \mathcal{T}_c$, $\mathcal{T} \leq \mathcal{T}_c$ by the application of general rules of thermodynamics. Since we are interested in the scaling of the spontaneous magnetization $\mathcal{M}$, we use

$$\mathcal{M} = \frac{\partial f}{\partial h}, \; at\; h = 0 \tag{9.34}$$

And obtain

$$\mathcal{M} = const.(\mathcal{T}_c - \mathcal{T})^{\frac{d-y_h}{y_t}} \tag{9.35}$$

A comparison with experimentally found law (9.3) yields

$$\beta = \frac{d - y_h}{y_t} \tag{9.36}$$

But also other physical quantities can be observed.

Specific heat

$$\frac{\partial^2 f}{\partial T^2}|_{h=0} = Const.(T_c - T)^{\frac{d}{y_t}-2} \tag{9.37}$$

So that in analogy to (9.36)

$$\alpha = 2 - \frac{\alpha}{y_t} \tag{9.38}$$

The evaluation of the susceptibility

$$\frac{\partial^2 f}{\partial h^2}|_{h=0} \tag{9.39}$$

Yields

$$\gamma = \frac{2y_t - d}{yt} \tag{9.40}$$

The importance of these results is that α, β, γ are determined by only *two* values y_t and y_h so that relations among α, β, γ exist, e.g.

$$\alpha + 2\beta + \gamma = 2 \tag{9.41}$$

All the above consideration exemplified by the Ising model of ferromagnetism can be extended to a whole class of PTs so that we have a *universality class* of scaling laws.

Concluding Remarks. A Lesson for Models in Urbanism

In retrospective we recognize a strange dichotomy of PT theories: highly sophisticated mathematical approaches (not presented here) on one hand, and on the other hand, a set of plausible model assumptions in conjunction with few-basic concepts. Both approaches yield experimentally verifiable, equivalent results. In view of the highly complex problem in urbanism it seems advisable to prefer the second kind of approaches.

But what is still more: this approach establishes quite specific relations between macroscopic "indicators" (quantifiers) such as temperature, magnetization, and specific heat based on processes on the microscopic level in such a way that the "fine structure", i.e. the details are no more important (a general property of systems close to a phase transition).

Obviously, such kind of approach will be highly desirable in urbanism. Our physics example shows that it may be possible. On the other hand, it is not obvious in how far the physics' approach can directly be transferred to urbanism. The self-similarity of street networks, power supply and Telecommunication networks etc. might be useful, but we must bear in mind that the "scaling process (cf. Fig. 9.4)" can comprise only three or four steps.

9.5 Nonequilibrium Phase Transitions (NPT)

In Sect. 9.3 of this chapter we have presented a few examples of NPTs. They can theoretically be treated by the formalism of Chap. 5 as follows. We consider a dynamical system represented by a state vector q with $\mathcal{J}$ components, where $\mathcal{J}$ may be very large. The system is subject to fixed conditions quantified by a set of control parameters Λ. The temporal evolution of the state vector, including its steady state, is determined by the evolution Eq. (5.11) of Chap. 5 with a purely deterministic part $\mathcal{N}(\Lambda, q)$ and a fluctuating part $F(t)$.

In a first step we ignore $F(t)$ and assume that for a fixed set of control parameters Λ there exists a stable steady state q_0. Then we change a control parameter such that q_0 becomes unstable which is checked by linear stability theory. We have assumed that only one real characteristic value λ_1 of the linearized equations becomes positive. It is connected with a characteristic vector q^1. Now we can take account of the nonlinearities and the fluctuating forces as outlined in Sect. 5.1 [Eqs. (5.25)–(5.27)]. While these equation can be solved by an iteration procedure, in many cases of practical interest (time-scale separation!) it suffices to consider the effect of the nonlinearities $\mathcal{N}(\Lambda, q)$ only in lowest order.

Then (5.26) reduces (in vector notation) to

$$q(t) = q^0 + \xi_1(t)q^1$$

where the amplitude $\xi(t)$ is the order parameter that obeys (5.28) or (5.29) depending on the degree of the nonlinearity of $\mathcal{N}$. Equations (5.28) and (5.29) are the relevant order parameter equations of the nonequilibrium phase transition. In the case of the laser the validity of (5.29) has been experimentally verified with high precision. In general, the dependence of $\mathcal{F}(t)$ on the original variable q is determined by the kind of system. Below we will study the case (5.28) in detail in connection with city dynamics.

For the moment being we deal with (5.29) as a prototypical equation of NPT. We assume that the fluctuating force $\mathcal{F}(t)$ obeys the Eq. (5.9), Sect. 5.1.

Thus the corresponding Fokker-Planck equation (5.44) and its steady state solution (5.46), both Sect. 5.3, can be established. The explicit form of (5.46), Sect. 5.3, is of particular interest. It reads ($\mathcal{N}$: normalization)

$$\mathcal{P}(\xi) = N \exp\left(-\frac{2\mathcal{V}(q)}{\mathcal{Q}}\right) \tag{9.42}$$

where

$$\mathcal{V}(q) = -\frac{1}{2}\lambda\xi^2 + \frac{b}{4}\xi^4 \tag{9.43}$$

There is a striking *formal* analogy between (9.42) and (9.43) and the probability $p(\mathcal{M})$ of the Landau theory of phase transition, where according to Sect. 9.1.

$$p(\mathcal{M}) = \mathcal{N}exp\left(-\frac{\mathcal{F}(\mathcal{M})}{\mathcal{RT}}\right) \tag{9.44}$$

where

$$\mathcal{F}(\mathcal{M}) = -\alpha(\mathcal{T}_c - \mathcal{T})\mathcal{M}^2 + \beta\mathcal{M}^4 \tag{9.45}$$

Both P(ξ) and $p(\mathcal{M})$ depend on ξ and $\mathcal{M}$, respectively, in the same manner, so that ξ corresponds to the *order parameter* $\mathcal{M}$ (in Landau's terminology). Furthermore, the noise strength $\mathcal{Q}$ corresponds to $k\mathcal{T}$, and most importantly, the control parameter λ corresponds to $\alpha(\mathcal{T}_c - \mathcal{T})$. When these parameters change their sign, the behavior of both systems, laser and ferromagnet, changes macroscopically and qualitatively. In spite of this beautiful analogy there are fundamental differences between (9.42) and (9.43) and (9.44) and (9.45) that are quite often overlooked in the literature. First, we mention some obvious and generally acknowledged differences.

(1a) the ferromagnet is in thermal equilibrium.
(1b) the laser in far away from it.
(2a) the result (9.43) can be derived from a microscopic theory.
(2b) the result (9.45) stems from a macroscopic, phenomenological approach.

The fundamental difference is less obvious: the quantities/parameters that determine (9.45) are of thermodynamic nature: temperature and energies. Whereas the quantities/parameters that determine (9.43) are rate constants such as production rate of light waves, their amplification and loss rates. Thus what we call "potential" $\mathcal{V}$ in (9.42) is not a "free energy" and cannot be calculated using energetic considerations. Having this caveat in mind, the "comeback" of the Landau theory may become a useful tool for urbanism: the phenomenological modeling of NPTs in cities. For instance, Landau considered also free energies of the form

$$\mathcal{F} = \cdot a\mathcal{M}^2 + c\mathcal{M}^3 + b\mathcal{M}^4 \tag{9.46}$$

Which we could formally "translate" into

$$\mathcal{V}(\xi) = -\frac{\lambda}{a}\xi^2 + c\xi^3 + b\xi^4 \tag{9.47}$$

The expressions (9.43) and (9.47) allow us to derive a scaling law of the order parameter ξ close to $\lambda = 0$, $\lambda > 0$. As an example we consider (9.43). The maxima of $\mathcal{V}(\xi)$ are at

$$\xi = \pm\xi_0 = \pm\left(\frac{\lambda}{4b}\right)^{\frac{1}{2}} \tag{9.48}$$

We choose as example the + sign and approximate

$$\mathcal{V}(\xi) \approx \mathcal{V}(\xi_0) + \lambda(\xi - \xi_0)^2 \tag{9.49}$$

Since—$\mathcal{V}$ appears as the exponent of an exponential function the essential contributions to it stem from a range

$$\lambda(\xi - \xi_0) \approx 1 \tag{9.50}$$

From this relation we derive the scaling law

$$(\xi - \xi_0) \propto \lambda^{-\frac{1}{2}} \tag{9.51}$$

When we debatably (s. below) ignore fluctuations, we may apply *bifurcation theory*. For a text cf. Luo (1997).

As in Sect. 5.1 we start from Eq. (5.11) but without $F(t)$ and scrutinize the system's behavior when due to the change of one (or several) control parameter (s) a steady state becomes unstable. We assume that linear stability theory applies. In a first step to *classify* bifurcation *scenarios* we use the number of eigenvectors that become unstable.

In the literature, typically their number is 1, 2, or 3, in exceptional cases also higher.

The analysis is confined to a region where the real parts of the eigenvalues are close to zero, so that the order parameters are small and the nonlinear function in their equation can be approximated by polynomials of low order.

Some typical examples are

(1) Only one eigenvector becomes unstable. Let the order parameter equation be

$$\dot{\xi} = \lambda\xi - b\xi^3 \tag{9.52}$$

In the steady state $\dot{\xi} = 0$ so that

$$\lambda\xi - b\xi^3 = 0 \tag{9.53}$$

This equation has three solutions

$$\xi_1 = 0, \xi_{2,3} = \pm\left(\frac{\lambda}{b}\right)^{1/2} \tag{9.54}$$

When we plot ξ against the "bifurcation parameter λ", the resulting diagram is shown in Fig. 9.7. Because of this shape it is called a pitchfork bifurcation.

(2) two eigenvectors become unstable

This may happen if

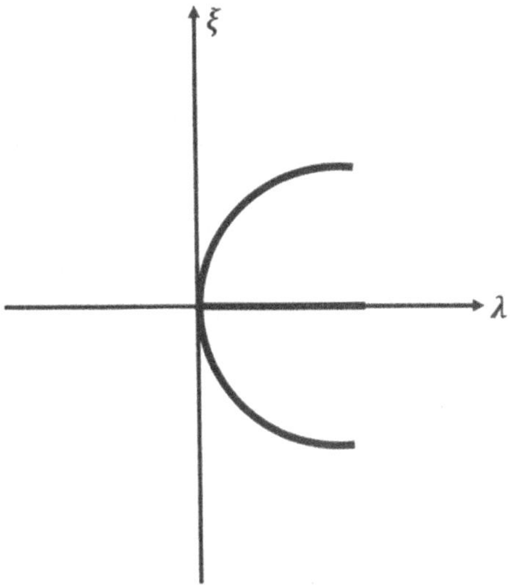

Fig. 9.7 Pitchfork bifurcation. OP ξ versus CP λ

(a) their real eigenvalues $\lambda_1 = \lambda_2$ coincide and become positive.
(b) if one eigenvalue is complex,$\lambda = \lambda_r + i\lambda_i$, and $\lambda_r \geq 0$.

λ_r, λ_i are the real and imaginary parts of λ.

(a) Typical OP-equations may read ($a > 0$, $b > 0$)

$$\dot{\xi}_1 = \lambda\xi_1 - \xi_1\left(a\xi_1^2 + b\xi_2^2\right) \tag{9.55}$$

$$\dot{\xi}_2 = \lambda\xi_2 - \xi_2\left(b\xi_1^2 + a\xi_2^2\right) \tag{9.56}$$

The steady state solutions are

$$\xi_1 = \xi_2 = 0 \tag{9.57}$$

$$\xi_1 = \pm\xi_2 \tag{9.58}$$

for $a \neq b$

(b) here the OP ξ is complex. Its complex conjugate is denoted by ξ^*.

The OP equation may read

$$\dot{\xi} = (\lambda + i\omega)\xi - b\xi\xi\xi^* \tag{9.59}$$

where $\lambda = \lambda_r$, $\omega = \lambda_i$; b is assumed real. By means of $\xi = re^{i\varpi t}$, $r \geq 0$ we obtain an equation for the real amplitude

$$\dot{r} = \lambda r - r^3 \tag{9.60}$$

Which has a pitchfork diagram reduced to $r \geq 0$.

This type of bifurcation with $\lambda_i \neq 0$ is called Hopf-bifurcation.

(3) when three eigenmodes become unstable the corresponding OPs may exhibit three types of behavior depending on the system's parameters

(a) Steady states
(b) Hopf-bifurcation
(c) Deterministic chaos

Its treatment is beyond the scope of our book.

Our examples are just a small selection of bifurcation scenarios. Just for illustration concerning case (2a).

Another scenario results when Eqs. (9.55) and (9.56) are slightly altered.

We put $a \equiv b$ and add a term $c\xi_2^2$ on the r.h.s of (9.56). Then in the steady state $\xi_2 = 0$ and ξ_1 undergoes a pitchfork bifurcation.

While formerly with $\xi_1 = \pm\xi_2$ the OPs coexisted, now they compete.

In conclusion, our view on the role of fluctuation. A look at Fig. 9.7 reveals the fundamental problem: There are three *possible* branches, but which one is chosen by a *real* system? This "choice" is made by the always present fluctuations. Thus their inclusion in any NPT-theory is inevitable.

References

Batty, M., & Longley, P. (1994). *Fractal cities*. London: Academic Press.

Bettencourt, L. M. A., Lobo, J., Helbing, D., Kühnert, C., & West, G. B. (2007). Growth, innovation, scaling, and the pace of life in cities. *Proceedings of National Academy Science U.S.A., 104*(17), 7301–7306. https://doi.org/10.1073/pnas.0610172104.

Cardy, J. (1996). *Scaling and renormalization in statistical physics*. Cambridge Univ press.

Dehaene, S. (2014) Consciousness and the brain. New York, NY, USA: Viking Press.

Frankhauser, P. (1991). Aspects fractals des structures urbaines. *L'Espace Geographique*, 45–69.

Gregory, R. (1970). "The intelligent eye" McGraw-Hill, New York (Photographer: Ronald C James).

Haken, H. (1996). *Principles of brain functioning: A synergetic approach to brain activity, behavior and cognition*. Berlin/Heidelberg/New York: Springer.

Landau, L. D., & Lifshitz, E. M. (1980). Statistical Physics. Vol. 5 (3rd ed.). Butterworth-Heinemann. ISBN 978-0-7506-3372-7.

Luo D. (1997) *Bifurcation theory and methods of dynamical systems*. World Scientific Singapore.

West, G. B., Brown, J. H., & Enquist, B. J. (1997). A general model for the origin of allometric scaling laws in biology. *Science, 276*(5309), 122–126. https://doi.org/10.1126/science.276.5309.122.

Wilson, K. (1983). The renormalization group and critical phenomena. *Reviews of Modern Physics, 55*(3), 583–600.

Chapter 10
Phase Transition and the City

10.1 Introduction

Phase transition (PT) and the city' is a very wide issue. At a grand macro scale it might refer to the PT of the urban revolution—the first appearance of urban society some 5500 years ago, while at a micro scale to the PTs in the life path of a single citizen: living with parents, at the city and at the suburb. In between, at the mezzo scale, there is still a wide range of scales such as PT processes of the emergence of cities and urban society out of Middle Ages feudalism, or still at a smaller scale, nowadays suburbanization and gentrification, or more specific case studies such as the balconies, lofts,.. and much more. So a chapter (not a book) on PT and the city must start with some definition of boundaries.

One way to set boundaries is to choose scale and specific case studies and start from there. This is what we'll do in the present chapter. We'll start (Sect. 10.2) with several case studies: the case of Tel Aviv balconies (10.2.1) which we've already described in brief in Chap. 4 (Sect. 4.4.2). Next (Sects. 10.2.2–10.2.5) we portray a scenario of the evolution of a metropolitan area. The scenario roughly follows the evolution of the Tel Aviv metropolitan area; yet with some modifications it can typify similar processes in other parts of the world. Central to our synergetics' interpretation of the above case studies is an interplay between control parameter (CP) and order parameter (OP). Thus, to generalize the case studies discussed in Sect. 10.2, we elaborate in Sect. 10.3 this issue by discussing the CP-OP-PT interplay in the context of potential urban scenarios. Based on the latter generalization we develop two mathematical models that illustrate the relations between CP,OP, PT: A model of 'growing cities' which represents the most typical urban process (Sect. 10.4), and a model referring to the information production process (Sect. 10.5). We conclude the chapter (Sect. 10.6) by an overview on the descriptive and mathematical parts of this chapter and their implications.

H. Haken and J. Portugali, *Synergetic Cities: Information, Steady State and Phase Transition*, Springer Series in Synergetics,
https://doi.org/10.1007/978-3-030-63457-5_10

10.2 Case Studies from Tel Aviv and Its Metropolitan Area

10.2.1 Tel Aviv Balconies

We've referred to the story of TA balconies in brief already in Chap. 4; here is "the full story". Open balconies were a dominant feature in the urban landscape of Tel Aviv from its day one at the beginning of the twentieth century. Influenced specifically by the modernist Bauhaus school and the local climate, almost all residential buildings were built with open balconies. During the 1930s and 1940s there were here and there attempts to close balconies and transform them to half rooms. These attempts, which implied a violation of the Tel Aviv planning law (and OP), were prevented by the city's planning authorities ("enslaved" in the language of synergetics). Yet the closed balcony possibility *was not forgotten.*

Following independence and massive immigration waves in the 1950 and the resulting shortage of residential space (CP), the closing balcony solution (OP) once again appeared (as "fluctuations") and in building after building balconies were closed (slaving diffusion). This time however, the municipality could not stop the process despite the fact that it was as before a violation of the planning law.

As noted in Chap. 4 (Sect. 4.4.2), the process started probably at end 1950s, when an unknown resident of Tel Aviv enlarged his/her apartment by closing the balcony and making it a "half-room" (Fig. 10.1, Balconies 1 and 2a). Some neighbors liked the idea and did the same. An interpersonal sequential SIRNIA design process gave rise to a space-time diffusion of closing balconies. As more and more balconies were closed, the form of the closed balconies gradually changed (Fig. 10.1). This process is still evolving as the remaining open balconies are being closed and balconies

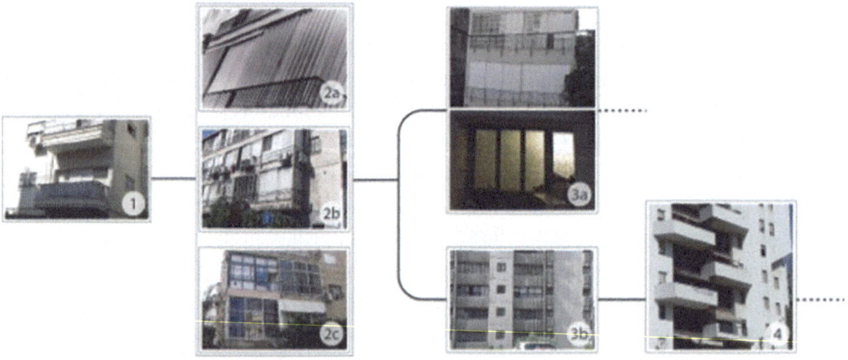

Fig. 10.1 "The butterfly effect of Tel Aviv balconies": 1 Open balcony. 2a Closed by asbestos shutters. 2b Closed by plastic shutters. 2c Closed by glass windows. 3a From the outside it looks as a balcony; from the inside part of the living room or a kitchen. 3b No balconies. 4 "Jumping balconies". Source: Portugali and Stolk (2014)

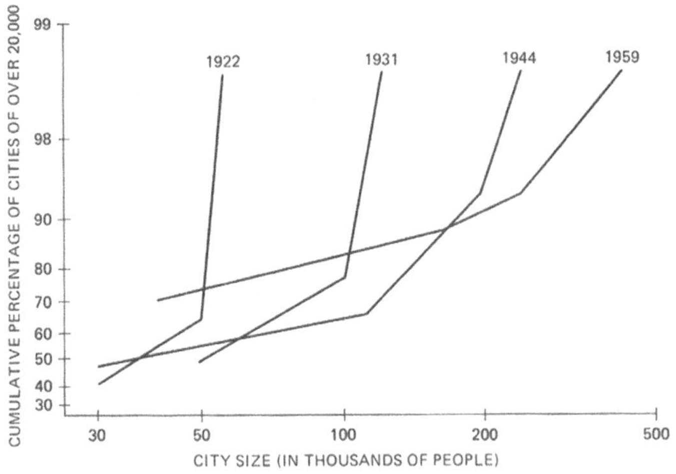

Fig. 10.2 Changes in the city size distribution of Israel 1922–1959. Source: Bell (1962)

previously closed in the older style are being renovated in line with the new more fashionable style.

The above, however, is not the end of the story. At a certain stage, the municipal planning authorities decided to intervene and started to tax all balconies, open and closed, as if they are a regular room. The response was another phase transition and a bifurcation: On the one hand, in old buildings, inhabitants continued closing balconies as above; on the other, developers with their designers started to design and build new buildings with already closed balconies, that is, with no balconies at all (Fig. 10.1).

Yet another small-scale phase transition took place at the end of the twentieth century, with the arrival of postmodern architecture: "Quoting" past patterns became fashionable and architects started to apply for permits to build balconies. Remembering their past planning experience, but wishing not to lag behind the advancing (post)modern style, the city planning authorities gave architects and developers permits to build open balconies, but in a way that, technically, would not enable them to be closed as in the past. The result was the "jumping balconies" which are typical to Israel's urban landscape only (Fig. 10.1). The very final stage and phase transition that happened some 15 years ago, was a return to square one: open balconies are once again allowed; closing them is prohibited.

Note' firstly, the role of memory, namely, that the closed balcony possibility was not forgotten. Secondly, that the phase transition of the first closed balconies was followed by a subsequent sequence of phase transitions reminiscent "aftershocks" that follow a major earth quake.

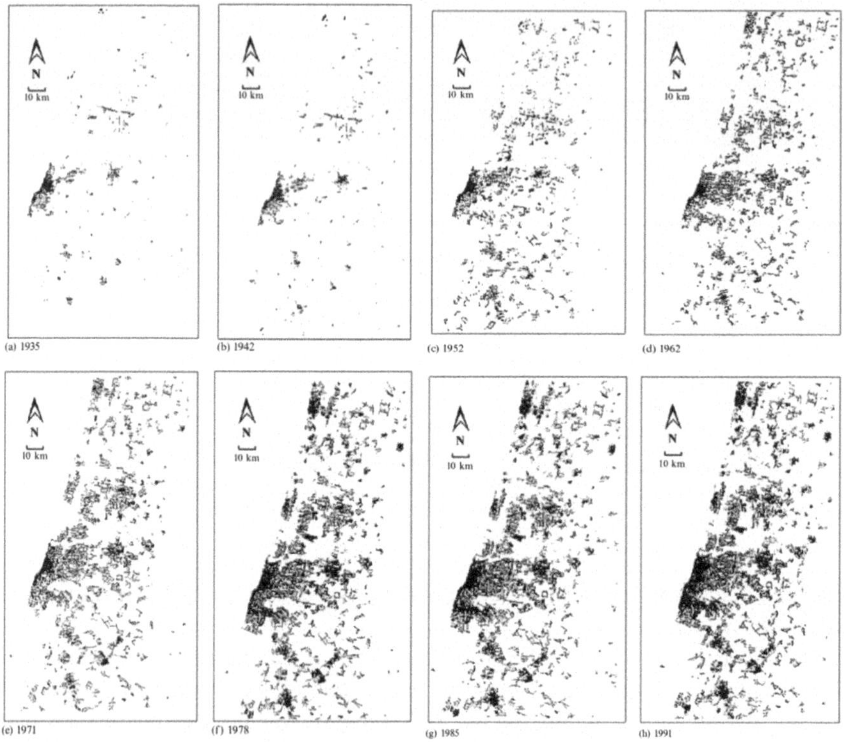

Fig. 10.3 The evolution of the metropolitan area of Tel Aviv. *Top from left to right*: 1935, 1942, 1952, 1962. *Bottom from left to right:* 1971, 1978, 1985, 1991. Prepared by Portugali

10.2.2 From Primate to Power Law Distribution

The above PT of the balconies and its subsequent events was an urban landscape expression to a mega-scale PT that following the war of independence and has dramatically shaken the Israeli urban system. While there is no room here to describe the process in details (see Portugali 1993), the following can be said: Following independence (in 1948), in less than 10 years the rank-size distribution of the Israeli urban system was transformed from a non-hierarchical *primate city* structure,[1] into a hierarchical power-law structure (Fig. 10.2).

The above transformation can be further observed in the sequence of maps of Fig. 10.3 that concentrate on the evolution of the Tel Aviv metropolitan area. Here too we observe a major structural change (PT) between the map of 1942 to that of 1952. The force behind this change (CP or OP) is the massive immigration waves that started in the late 1940s, before the establishment of the state of Israel, of Jews

[1] One/few big cities and many small settlements with no intermediate-size cities in between (Jefferson 1939).

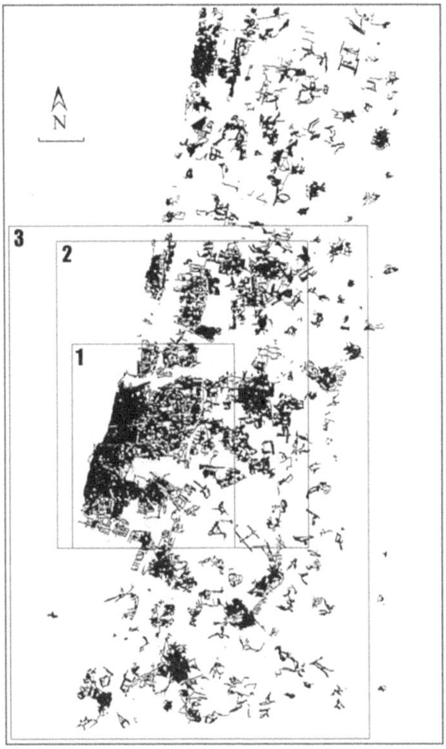

Fig. 10.4 Map of the entire Tel Aviv metropolis with three study regions: 1 central part; 2 northern part; 3 entire ensemble (After Benguigui et al. 2000)

from Europe who immigrated (illegally) to Israel that was then under the British mandate. After independence, the waves intensified, with immigration waves of Jews from Europe and from Islamic countries, specifically from Yemen (1948–1950), Iraq (1950–52) and North Africa (mid 1950s). In less than 10 years, these waves have tripled the population of Israel as a whole and specifically its central area, north and south of Tel Aviv. Temporary immigrants tents camps were transformed into towns and previously relatively independent small towns and agricultural settlements, experienced a sudden and fast population growth.

10.2.3 Suburbanization and Gentrification

The growth of the urban system continued from the 1970s onwards, but this time not due to immigration waves, but as a consequence of the "classical" suburbanization-metropolinization internal migration processes. Since end 1960s and in the 1970s, Tel Aviv was once again subject to housing shortage, this time, however, due to "classical" urban processes that were typical of many cities around the world: natural demographic growth, rising standards of living, high rate of car ownership, people in old residential neighborhood get older, young families move out to nearby towns but continue to work in Tel Aviv, the number of Tel Aviv residents is declining, previous residential neighborhood are transformed to CBD (central business districts) … Tel Aviv becomes the metropolitan core, firstly to suburban towns around it and a bit later to a larger area—to agricultural rural settlements thus transforming them to suburbs of the big city (a process often termed as *rurbanization*). The process as a whole is well represented by the set of maps in Fig. 10.3.

In the end 1970s and during the 1980s a new PT and OP—*Gentrification*: "Yuppie" ("young urban professional") and then "regular" young middle-class families moved from the suburbs back to the city centers, pushing out the poor and occupying their neighborhoods, as well as previously residential areas that have been transformed to CBD or semi-industrial areas at the city center (e.g. lofts) and now once again became residential. The latter processes of gentrification entailed, firstly, *The Rise of the Creative Class* (Florida 2002) that plays a key role in the socio-economy of today's post-industrial cities—specifically in global cities that are connected to the global socio-economy. Secondly, to a socio-economic and cultural gap between the central global cities and the peripheral towns and cities that "were left behind"—a gap that shows itself in the Brexit in England, the Yellow jacket in France and more. In both suburbanization and gentrification, the PT gave rise to a new OP and to a long period of steady state (e.g. commuting in the first and walking/cycling etc. in the second).

10.2.4 The Evolving Fractal Structure

It is interesting to note that the above geo-historical description of the Tel Aviv metropolitan area shows up in the evolution of the fractal dimension of that area. Based on the set of maps in Fig. 10.3, Benguigui et al. (2000) have studied the time evolution of the fractal dimension of the region that now forms the metropolitan area of Tel Aviv. The study started by identifying, by means of visual inspection of the maps in Fig. 10.3, three study regions in the metropolitan area as presented in Fig. 10.4.

The study suggests that the central part of the Tel Aviv metropolitan area, that included Tel Aviv and its adjacent towns (region 1 in Fig. 10.4), was from the start fractal, when its fractal dimension D increased from 1.533 in the year 1935, to 1.809 in 1991, with a rapid increase between 1971 and 1978; apparently due to the suburbanization process noted above. Region 2 too was from the start (1935) fractal, however, with D values ranging from 1.387 in 1935 to 1.733 in 1991, with a pronounced "jump" between the years 1941 and 1952, reflecting the immigration waves before and after 1950 noted above. The gap in the fractal dimensions between regions 1 and 2 indicates that the Tel Aviv metropolitan area is not yet fully integrated and unified. Finally, judging from the evolving D of region 3 one can say that the entire metropolitan area of Tel Aviv became fractal only after 1985.

10.2.5 Intermediate Discussion

As in the story of the balconies, so in the process of migration to, and urbanization of, the agricultural villages into suburb towns, the process started as SIRNIA (Chap. 4) derived fluctuations in the form of violation of the planning law. In the case of the agricultural villages, the planning law (OP) prohibited to transform agricultural land into built-up area. In fact, what we had here, was a competition between two OPs: an *urban OP* driven by demographic growth, housing shortage, increasing demand for, and prices of, urban land; versus an *agricultural OP* driven by *The Governmental Committee for the Preservation of Agricultural Land*, itself driven by the Zionist ideology "to return to the land". As implied from the above, the urban OP ("market forces") won the competition: many farmers/land owners found it beneficial to sell parts of their land to urban private developers. And again, as in the case of balconies, after a decade or two the authorities had no choice but to legalize the transformation.

In the case studies described above, we can identify three major PTs: the case of the Tel Aviv *balconies*, the process of *suburbanization* and the counter process of *gentrification*. In all three the PT took the form of a space-time diffusion process composed of a sequence of smaller-scale PTs. In the case of the closed balconies PT, the closed balconies were followed by buildings with no balconies, then by "jumping balconies" and finally, once again, by open balconies. The suburbanization PT was followed, or was associated with, a sequence of smaller scale PTs: the transformation of Tel Aviv into a CBD and in parallel, the urbanization of the rural-agricultural settlements ("rurbanization"). In each individual agricultural settlement, such a change implied a local PT. Finally, in the case of the gentrification PT, the subsequent sequence of smaller scale PTs included the transformation of Tel Aviv into a mixed uses residential ("creative class") + "smart" CBD.

Thus, unlike PTs in physics (laser, ice to water,...) where due to the slaving principle an emerging OP is followed by StS, in cities, the PTs are followed by slaving that takes the form of a sequence of small-scale PTs and further events and developments. In this respect the OP is similar to what Bohm and Peat (1987) have termed "generative order", an order that has the potential to generate other orders or events.

10.3 The Interplay Between OPs and CPs in PT

As we've seen in the case studies just described, central to them is a play between population (e.g. numbers of immigrations and/or natural increase) and housing (e.g. availability of flats). As we've seen in Chap. 3 above and in subsequent chapters, central to Synergetics is the interplay between control parameter (CP) and order parameter (OP). This raises the question 'who is how?'—which of the two is OP and which CP?

Table. 10.1 Urban scenarios and their variables

Scenarios	q: OP	λ: CP	γ: saturation rate	Q: immigration/emigration import rate
Growing cities	Citizens	Flats, jobs, quality of life	Exhaustion of facilities	Taking residence, immigration
Shrinking cities	Flats, jobs	Population		Emigration
New products	New products	Citizens, producers, buyers	Market saturation	Import
Coronavirus	Infected people	Infection rate	"Herd immunity"	Immigration of infected people
Information production (IP)	Information	Citizens, artifacts	Citizens' capacity	Information from outside

An answer to this question comes from Synergetics 1st Foundation, cf. e.g. Chap. 3. There is a unique criterion based on time-scale separation. CPs are constant or change at least much more slowly than OPs. Thus housing facilities that change at least in general much more slowly than population size are CPs and population size OP. On the other hand, a technical innovation that spreads/grows much more rapidly than population size is now the OP. The last decades events in Detroit US present yet another example: here, a drastic drop in a city's economy, entailed massive outmigration[2] and as a consequence, a "wave" of surplus flats and houses. As a result, the relations between population and housing were transformed: the number of available flats and houses became OP that was "controlled" by population as CP, as is evident by the fact that many houses were blocked. A PT thus occurs when a CP is changed beyond a critical value and the OP has to respond to this change or even a new OP is generated because of some innovation. Thus PTs are singular situations.

The relations between CP, OP, and by implication PT, depends as we've just seen on the specific circumstances that take place in a city or system of cities. As noted above, in all cases *the decisive distinction between OPs and CPs rests on the criterion of time-scale separation*. In Table 10.1 we provide a list of typical urban scenarios and their entailed CP, OP relations. We describe each scenario by four variables that we use in the mathematical models we develop in Sect. 10.4 below: OP (q), CP (λ), saturation rate (γ), and immigration rate (Q). As will transpire below, these variables are associated with the Verhulst equation, also known as "logistic equation" which has found a number of applications in various disciplines. In the model developed below we use this equation in a novel way by adding to it a random "force" F.

Growing cities. This is the most typical urban development in most parts of the world: urban population and thus cities are growing very fast. In Europe, due to immigration, in the Far East due to massive migration processes from the rural countryside to the urban and metropolitan centers: In India and other countries, as

[2]The city population declined from 1,850,000 in 1950 to 680,000 in 2015 and 672,662 in 2020.

a consequence of spontaneous, self-organized processes, while in China as part of centralized governmental policy. The result of these processes is an unprecedented reality in which for the first time in human history more than half of world population lives in cities.

Shrinking cities. Also called *counter-urbanization*, referring to cities who are losing population for various reasons ranging from economic crisis to "regular' migration from peripheral rural regions to metropolitan centers (Pallagst et al. 2014). In some cases shrinking is associated with socio-economic crises, as recently in Detroit, while in others shrinking cities become even more prosperous than before (Hartt 2018).

New products. This refers to the "classical" spatial diffusion processes due to new products, innovations, and so on (see detailed discussion in Chap. 11 that follows). Just consider the effect the invention of a new product such as the 'automobile' had on cities—on their structure, urban landscape, daily life and much more. Following what Schwab (2016) has termed *The 4th Industrial Revolution*—the so called Industry 4.0 with its smart artifacts—there are currently *smart cities* studies attempting to appreciate the effect of smart devices on cities. In Chap. 14 we deal with this issue in some details.

Coronavirus. Similar to the above, with virus instead of innovation. The coronavirus has all the ingredients of a complex SO system: unpredictability, uncertainty, abruptly emerging PT and so on. A specifically interesting question concerns its future effects on several socio-economic and cultural trends: increasing economic globalization, on the one hand, versus increasing nationalist feelings on the other (e.g. Brexit). One lesson from Synergetics is the effect and role of fluctuations (Chaps. 3, 4) is that when they occur in instable periods their effect might be dramatic. See further notes in the concluding chapter.

Information production (IP). As noted in Chap. 4, "Urban dynamics is a kind of production process—producing artifacts of all kinds ... These artifacts convey data from which urban agents extract SI and PI with their entailed SHI; .... We term this SIRN-IA process the City's *information production* (IP). This IP process can be seen as, or gives rise to, the city's order parameter ...".

Of the above five scenarios, the first—growing cities—is the most typical urban process, while the fifth—IP is the more general and can thus be applied to each of the other scenarios as well as to specific economic, social or cultural events that take place in a city. In what follows we thus develop two models: one regarding the case of growing cities and one regarding IP.

10.4 Growing Cities

In this section we deal with population dynamics, where we assume that the number of citizens q plays the role of OP. Later we will discuss other interpretations of q. We base our approach on the results of Chap. 5 where we derived a prototypical OP equation for numbers, i.e. Equation (5.28). By an obvious change of notations, we write Eq. (5.28) in the form

$$\frac{dq}{dt} = \lambda q - \gamma q^2 + F(q, t) \tag{10.1}$$

When we put $F = 0$, (10.1) becomes the familiar Verhulst equation. λ and γ play the role of control parameters (CP) which we consider as given and time-independent. The important new feature of (10.1) as compared to the Verhulst equation is the occurrence of fluctuations/random force $F(q, t)$. As we have seen in Chap. 9, such forces play an important role in PTs and NPTs. The specific choice of the form of $F(q, t)$ must be based on the situation considered. Here we consider the case that the number of citizens may randomly change due to a change of residence—people moving into or out of the city as described in Sects. 10.1 and 10.2 above. As we've further seen above, these moves may happen at a low rate or a high one—an immigration wave. In order not to load the formalism, we treat only moves into the city. Since the mathematical details are somewhat involved (and not essential for our presentation) we sketch only the essential steps. We require that the fluctuating force $F(q, t)$ obeys a relation of the form

$$\langle F(t)F(t^{'})\rangle = q(t)Q\delta(t - t^{'}) \tag{10.2}$$

The brackets $\langle \ldots \rangle$ indicate the average over the random process. Q is the fluctuation strength and $\delta(t - t^{'})$ Dirac's function. Note the occurrence of $q(t)$ in (10.2) which means that the size of the random fluctuations is proportional to the population size. This implies mathematical intricacy. Since the population size q changes with each move abruptly, the value of q to be used remains undefined.

Here we use the Stratonovich calculus according to which we use the average value of q before and after the move. Under this assumption we may derive a Fokker-Planck equation for the distribution function $f(q; t)$ (for some more details cf. 10.4.1 below).

$$\dot{f} = -\frac{\partial(Kf)}{\partial q} + \frac{Q}{4}\frac{\partial f}{\partial q} + \frac{Q}{2}\frac{\partial}{\partial q}(q\frac{\partial f}{2q}) \tag{10.3}$$

where

$$K = \lambda q - \gamma q^2 \tag{10.4}$$

In the following we want to calculate the time evolution of the mean value of q and its variance $S = \langle q^2\rangle - \langle q\rangle^2$ as well as their steady state values.

1. Time evolution of $\langle q\rangle$
   We multiply (10.3) by q and integrate both sides of q from $q = 0$ till $q = \infty$; using the definition

$$\int_0^\infty qf(q,t)dq = \langle q(t)\rangle \equiv \langle q\rangle \tag{10.5}$$

and the normalization of f, $\int_0^\infty f dq = 1$ for all times we obtain

$$\frac{d}{dt}\langle q\rangle = \lambda\langle q\rangle - \gamma\langle q^2\rangle + \frac{Q}{4} \tag{10.6}$$

2. Time evolution of S

In an analogous way we obtain

$$\frac{dS}{dt} = 2\lambda S - 2\gamma\left(\langle q^3\rangle - \langle q\rangle\langle q^2\rangle\right) + 2Q\langle q\rangle \tag{10.7}$$

The last term in (10.6), $Q/4$, represents an average increase of population due to taking residence/immigration. This increase entails an increase of variance according to the last term in (10.7), $2Q\langle q\rangle$.

Because of nonlinear terms, such as $\langle q^2\rangle$ in (10.6) and (10.7), these equations can only approximately be solved (see below). On the other hand, we obtain an explicit solution of the Fokker-Planck equation (10.3) in the case of steady state where

$$\frac{df}{dt} = 0 \tag{10.8}$$

$$f(q) = Nq^{-1/2}\exp(\alpha q - \beta q^2)$$
$$\lambda = \frac{2\alpha}{Q}, \beta = \frac{\gamma}{Q} \tag{10.9}$$

Most importantly, the normalization constant N and all moments $\langle q^n\rangle$, $n = 1, 2, \ldots$ can be expressed by well-known integrals (see below). For our purpose, we can cast them in a handy form for important limited cases. In the spirit of PT theories we consider the transition from one steady state to a new steady state. In the following we consider several scenarios. We begin with the following situation. In the beginning, at time $t = 0$, we assume $q = 0$, i.e. no population. This is, of course, a very artificial assumption. But this model allows us to familiarize us with the general approach. In addition, it allows us to deal with realistic cases if we identify q with the number of a specific industrial product, e.g. electric cars.

We assume that the parameters λ γ, Q are fixed. To solve (10.6) approximately, we approximate

$$\langle q^2\rangle by \langle q\rangle^2 \tag{10.10}$$

which is a good approximation if the distribution function is sharply peaked which we are not sure of, however. Therefore it is important to be able to compare the approximate solution to (10.6) with an exact result.

Under the assumption (10.10), the steady state solution to (10.6) reads

$$\langle q\rangle = \frac{\lambda}{2\gamma} + \left(\frac{\lambda^2}{4\gamma^2} + \frac{Q}{4\gamma}\right)^{1/2} \tag{10.11}$$

We compare this with the exact values based on the exact solution of the Fokker-Planck equation (10.9) for the two limiting cases "λ small" and "λ large". In the case "λ small" we assume

$$\lambda^2 \ll Q\gamma \tag{10.12}$$

Then

$$\langle q\rangle \approx \frac{1}{2}\left(\frac{Q}{\gamma}\right)^2 \tag{10.13}$$

whereas the "exact" result reads (cf. below)

$$\langle q\rangle = 0.3\left(\frac{Q}{\gamma}\right)^{1/2} + \text{ small corrections.} \tag{10.14}$$

When "λ large", we neglect the term containing Q so that (10.11) becomes

$$\langle q\rangle \approx \frac{\lambda}{\gamma}$$

The exact result reads

$$\langle q\rangle = \frac{\lambda}{\gamma} + \text{ small corrections} \tag{10.15}$$

This remarkable coincidence allows us also to study the time-dependence of $\langle q\rangle$ and S based on Eqs. (10.6) and (10.7) under the approximation (10.10).

While the solutions can be found without any further approximations, we confine our analysis to the initial phase where the nonlinearities can be ignored. From (10.6) we obtain

$$\langle q(t)\rangle = \frac{Q}{4\lambda}(e^{\lambda t} - 1) \tag{10.16}$$

An interesting question is, at which time t the behavior (10.16) stops because of the effect of saturation. Since we may assume that the exact $\langle q(t)\rangle$ follows approximately a sigmoidal, i.e. S-shaped curve, we postulate that (10.16) comes close to a fraction, say ½, of the steady state value, be it approximate or exact. In the case "λ small" [cf. (10.12)] we find because of (10.16) and (10.13)

$$e^{\lambda t} - 1 \approx 2\frac{\lambda}{Q}\left(\frac{Q}{4\gamma}\right)^{\frac{1}{2}} = \lambda\frac{1}{(\gamma Q)^{1/2}} \ll 1 \tag{10.17}$$

or

$$e^{\lambda t} \ll 2 \tag{10.18}$$

which means a comparatively slow increase of $q(t)$.

On the other hand, if "λ large" we arrive at

$$e^{\lambda t} \gg 1 \tag{10.19}$$

i.e. an exponential increase of $q(t)$.

The solution to (10.7) allows us to discuss the behavior of the variance during the initial phase that corresponds to that of $\langle q(t)\rangle$ (10.16). Again we neglect the nonlinear terms. An elementary calculation, without any further approximation, leads us to

$$S(t) = \frac{Q^2}{4\lambda^2}(e^{\lambda t} - 1)^2 \tag{10.20}$$

or, in terms of $\langle q(t)\rangle$

$$S(t) = 4\langle q(t)\rangle^2 \tag{10.21}$$

We turn to our second scenario where we start from the steady state of a city with $q_0 > 0$ inhabitants. We study what happens to the population size q, if either the growth parameter λ or the fluctuation parameter Q are suddenly and appreciably increased. In the first case, we deal with a sudden improvement of living conditions, e.g. after the end of a war, end of an economic crisis etc. A case in point is the situation in Israeli cities after the 1967 Six Days War that was followed with an economic boom. In the second case, we may think of immigration waves, that as described above, in the early 1950s followed the independence of Israel. In principle, we may also think of a sudden decrease of the saturation parameter γ, e.g. by a sudden building boom, though this won't be treated here.

Our previous treatment of the scenario, where initially $q = 0$, allows us to deal with the present scenarios in an elegant fashion. Namely, the equations for $\langle q(t)\rangle$ and S (10.6) and (10.7), and the steady state $f(q)$ hold also now, provided we insert the new CPs and solve (10.6) and (10.7) under the new initial conditions where we use steady state values of $\langle q\rangle$, S evaluated with the old parameter values. Again we consider an initial transition phase where we may neglect the quadratic terms in (10.6) and (10.7).

We begin with scenario (2) where λ is increased $\lambda > \lambda_0$, where λ_0 former value, but Q is kept fixed. Since the expression $\langle q \rangle$ is somewhat clumsy, we introduce the denotation

$$X = \langle q \rangle \tag{10.22}$$

that allows us to simply indicate the needed attributes of X, namely initial state: X_i, final state: X_f, time dependence and parameter λ used: $X(t, \lambda)$. By use of these denotations, we may write the time-dependent solution to (10.6) (quadratic terms ignored)

$$X(t; \lambda) = \frac{Q}{4\lambda}(e^{\lambda t} - 1) + X_i(\lambda_0)e^{\lambda t} \tag{10.23}$$

In analogy to scenario (1) above, we estimate the time t after which X equals a fraction, say ½, of the final value $X_f(\lambda)$. We may either use the exact or the approximate values of X_i, X_f. We obtain

$$e^{\lambda t} \approx \left(\frac{Q}{4\lambda} + X_i(\lambda_0)\right)^{-1} (\frac{1}{2}X_f(\lambda) + \frac{Q}{4\lambda}) \tag{10.24}$$

We evaluate (10.24) in the case "λ_0 small", "λ large", which implies a considerable increase of λ_0. In this case

$$X_i(\lambda_0) \approx \frac{1}{2}\left(\frac{Q}{\gamma}\right)^{1/2} \text{ or "exact"}: 0.3\left(\frac{Q}{\gamma}\right) \tag{10.25}$$

$$X_f(\lambda) \approx \frac{\lambda}{\gamma} \text{ also"exact"} \tag{10.26}$$

Using the inequality "λ large", $\lambda^2 \gg Q\gamma$ we may readily evaluate the leading term on the r.h.s. of (10.24), so that

$$e^{\lambda t} \approx \frac{\lambda}{(Q\gamma)^{1/2}} \gg 1 \tag{10.27}$$

which implies a rapid increase of *X(t)*, i.e.$\langle q(t) \rangle$. As we may show, S increases similarly rapidly. To conclude scenario (2), i.e. increase of λ, we consider the case that both λ_0 and λ are "large" so that

$$X_i(\lambda_0) = \frac{\lambda_0}{\gamma}, X_f(\lambda) = \frac{\lambda}{\gamma} \tag{10.28}$$

Under the realistic assumption that

$$Q \ll \lambda \tag{10.29}$$

(possibly excluding the case of an immigration wave, see below), the relation (10.24) reduces to

$$e^{\lambda t} \approx \frac{1}{2}(\lambda/\lambda_0) \tag{10.30}$$

Putting $\lambda = \lambda_0 + \Delta\lambda$ and assuming

$$\frac{\Delta\lambda}{\lambda_0} < 1 \tag{10.31}$$

we may replace (10.30) by

$$\lambda t \approx \Delta\lambda_0 \tag{10.32}$$

which implies a moderate growth of $e^{\lambda t}$.

Our 3rd scenario deals with the impact of an immigration wave where Q is suddenly and considerably increased. Having the inequality (10.12) in mind, which we interpret in terms of Q, we consider the transition from Q_0 small, corresponding to "λ large" to Q large, corresponding to "λ small". We adopt the notation (10.22) and use (10.24) where X_i (Q_0) replaces $X_i(\lambda_0)$. If "λ large", then

$$X_i(Q_0) = \frac{\lambda}{\gamma} \tag{10.33}$$

whereas now, "λ small"

$$X_f(Q) \approx \frac{1}{2}(\frac{Q}{\gamma})^{1/2} \tag{10.34}$$

Under the provision that we use (10.33) and (10.34) in (10.24) we may evaluate (10.24)

$$e^{\lambda t} \approx 1 + \frac{\lambda}{(\gamma Q)^{1/2}} + \dots \tag{10.35}$$

or equivalently

$$t \approx \frac{\lambda}{(\gamma Q)^{\frac{1}{2}}} + \dots \tag{10.36}$$

This result is remarkable because the time t after which saturation sets in, does not depend on the replication rate λ, but only on the saturation coefficient γ (representing e.g. accommodation facilities) and the immigration influx proportional to Q. Note that this "observation" is relevant to the situation in Israel immediately following independence, as described above: the first response to the influx of new immigrants was to build "tent settlements" (Ma'abarot in Hebrew, meaning transition settlements). At a later stage they were transformed to "development towns". Note further that nowadays such "tent towns" are widespread phenomenon in many parts of the world to overcome "saturation limits".

10.4.1 Some Mathematical Details

Using a precise mathematical formulation we write Eq. (10.1) as stochastic differential equation

$$dq(t) = K(q(t))dt + g(q(t))dv(t) \tag{10.37}$$

where $\langle dv \rangle = 0, \quad \langle dv(t)dv(t) \rangle = dt.$

By use of Stratonovich calculus the Fokker-Planck-equation belonging to (10.37) reads

$$\frac{df}{dt} = -\frac{\partial(Kf)}{\partial q} + \frac{1}{2}\frac{\partial}{\partial q}\left(\left(g\frac{\partial}{\partial q}g\right)f + g^2\frac{\partial f}{\partial q}\right) \tag{10.38}$$

In the steady state

$$\frac{df}{dt} = 0 \tag{10.39}$$

This allows us to integrate (10.38) which yields, observing the integrability condition

$$\lim_{q\to\infty} f(q) = 0$$

$$-K(q)f + \frac{1}{2}\left(g\frac{\partial}{\partial q}g\right)f + \frac{1}{2}g^2\frac{\partial f}{\partial q} = 0 \tag{10.40}$$

To make contact with (10.1) with specify

$$K = \lambda q - \gamma q^2, \quad g = \sqrt{Q}\sqrt{q}, \quad q \geq 0 \tag{10.41}$$

Inserting (10.41) in (10.40) and a slight rearrangement of terms leads us to

$$\frac{\partial f}{\partial q} = \left(2Q^{-1}(\lambda - \gamma q) - \frac{1}{2q}\right) f \tag{10.42}$$

The solution to (10.42) reads

$$f(q) = Nq^{-1/2} \exp\left(\alpha q - \beta q^2\right) \tag{10.43}$$

with N normalization constant, and

$$\alpha = \frac{2\lambda}{Q}, \beta = \frac{\gamma}{Q} \tag{10.44}$$

The moments $\langle q^n \rangle$, $n = 1, 2, \ldots$ can be obtained by derivatives of the partition function

$$Z = \int_0^\infty q^{-1/2} \exp\left(\alpha q - \beta q^2\right) dq \tag{10.45}$$

e.g.

$$\langle q \rangle = \frac{\partial}{\partial \alpha} \ln Z \tag{10.46}$$

By means of the transformation.
$q = \xi^2$, Z can be cast into the form

$$Z = \int_{-\infty}^\infty \exp\left(\alpha \xi^2 - \beta \xi^4\right) d\xi \tag{10.47}$$

which has been exactly evaluated. In fact, (10.47) is a special case of the standard integral

$$\int_{-\infty}^\infty \xi^n \exp\left(\alpha \xi^{2m} - \beta \xi^{4m}\right) d\xi = (2m)^{-1} (2\beta)^{-(n+1)/4m} \Gamma\left(\frac{n+1}{2m}\right)$$
$$D_{-(n+1/2m)}(K) \exp\left(K^2/4\right),$$

where $K = -\alpha/\sqrt{2\beta}$, Γ : gamma function, D parabolic cylinder function, n, m integers > 0.

Equations (10.14) and (10.15) of the main text can be derived from (10.47) and (10.37) as follows. If λ or α [cf. (10.44)] is small, we may expand part of the exponential function in (10.47) in terms of powers of α (Taylor series). If λ is large, the integral in (10.47) can be evaluated by the method of steepest descent.

10.5 Information Production (IP)

10.5.1 Basic Model

In this Sect. we cast the results of Chap. 4 on SIRNIA, in particular on IP, in a quantitative form. We focus our attention on the production of Shannonian information (SHI) and pragmatic information (PI) and their interplay. We denote the quantity of SHI, averaged over 24 h, by s, and that of PI by p. Both s and p are measured in bits and refer to the whole city with n citizens. On average, each citizen has a certain capability of converting information into action. We denote the daily conversion rate by r. Then the total amount of PI measured by p increases per day (24 h) by

$$nrs \tag{10.48}$$

On the other hand, there is a daily decay of possibilities to perform actions for a variety of reasons. This decay is proportional to p and a rate γ.We assume that γ is a constant, i.e., independent of p and s.

Taking (10.48) and the losses $-\gamma p$ together, we arrive at the net production eq. for p

$$\frac{dp}{dt} = nrs - \gamma p \tag{10.49}$$

We turn to the production of SHI, i.e. s. Each citizen contributes by his/her action to an increase of s at a rate r, so that the total increase of s (per day) is given by

$$nr'p \tag{10.50}$$

Due to "collective forgetting" and other reasons there will be a loss of s at an average rate r that we assume to be constant.

Finally, there are events beyond the control of citizens. These events are of a stochastic nature and contribute to an increase of s (i.e. of uncertainty). We take their effect into account by a random force $F(t)$. Thus our 2nd equation reads

$$\frac{ds}{dt} = nr'p - \Gamma s + F(t) \tag{10.51}$$

Equations (10.49) and (10.51) form our basic model. In the following we explore its most interesting implications step by step.

10.5.2 *Information Explosion*

Since this effect exists whether or not there is a fluctuation force, in (10.51) we put $F = 0$. To solve (10.49) and (10.51) we put

$$p = p_0 e^{\lambda t}, s = s_0 e^{\lambda t} \tag{10.52}$$

and obtain, in the standard way,

$$\lambda = -\frac{\Gamma + \gamma}{2} \pm \frac{1}{2}\left((\Gamma - \gamma)^2 + 4n^2 r r'\right)^{1/2} \tag{10.53}$$

An instability occurs, if one characteristic value $\lambda > 0$. This happens if

$$n^2 r r' > \gamma \Gamma \tag{10.54}$$

Provided rr' and $\gamma\Gamma$ are constant, such an instability may happen if the number of citizens exceeds a critical value. One effect might even be the breakdown of the whole information system.

As Eq. (10.54) reveals, the instability can be avoided if we decrease the conversion rates r or/and r', or if we increase the loss rates γ or/and Γ. Below we will treat a realistic example in more detail.

10.5.3 *What is the OP?*

Since our model contains the two variables p and s, we have to invoke a criterion that allows us to identify the OP. To this end, we apply the time-scale separation criterion: The "long living" variable enslaves the "short living" one. But what means "long/short living"? In the context of information it is the *memory*. Clearly that of a society is much longer than that of an individual. Since SHI refers to the city in total, whereas PI refers to the action of individuals, clearly $\Gamma \ll \gamma$ so that s must be the relevant OP.

Within the formalism of synergetics, this allows us—in a good approximation to put in (10.49)

$$\frac{dp}{dt} \approx 0 \tag{10.55}$$

so that

$$p \approx \frac{nrs}{\gamma} \tag{10.56}$$

Inserting (10.56) into (10.51) we obtain

$$\frac{ds}{dt} = Rs + F(t) \tag{10.57}$$

where

$$R = \frac{n^2 rr'}{\gamma} - \Gamma \tag{10.58}$$

is the net replication rate.

$R > 0$ is equivalent to the instability condition (10.54). If R < 0, a stable, but stochastic, i.e. fluctuating, on average finite state of s is reached. In view of the actual, realistic situation with ever increasing values of p and s, this situation is rather unlikely, however. So we deal with the more interesting and important case of $R > 0$.

10.5.4 What May Curb the Unlimited Growth of s (and p)?

From a theoretical point of view this goal can be reached by a lowering of $R > 0$ to $R = 0$. To achieve this we may think of governmental regulations or "natural" mechanism. In the spirit of our book that deals with selforganization, we disregard governmental regulations (e.g. taxation of SHI production or transmission). Rather our focus is the behavior of citizens who convert SHI into action and is captured by the conversion coefficient r. It is most plausible that a citizen's capability of converting s into p decreases with increasing s. We model this effect by putting

$$r = r_0(1 + as)^{-1} \approx r_0 - bs \tag{10.59}$$

where $b > 0$ is constant.

By inserting our hypothesis into R(10.58) and thus in (10.59), we obtain after rearrangement of terms

$$\frac{ds}{dt} = as - Cs^2 + F(t) \tag{10.60}$$

where

$$a = \frac{n^2 r r'}{\gamma} - \Gamma, \quad C = \frac{n^2 b r'}{\gamma} \tag{10.61}$$

Equation (10.60) becomes identical with the generalized Verhulst equation (10.1) of Sect. 10.4, provided we subject the fluctuating force to the same conditions as F in (10.2). This is a very natural condition because it gives rise to a constant influx rate as we have shown in Sect. 10.4. Now, we may apply the whole analysis of Sect. 10.3 to the present case of IP provided we identify $\lambda \rightarrow a, \gamma \rightarrow C, Q$ as before.

10.5.5 Concluding Remark on Information Production

Based on plausibility arguments, we have derived equations for the production of Shannon and pragmatic information. Surely, in the future, more detailed theories will be required. Nevertheless, as an important and realistic result we obtained an unlimited growth of SHI and PI, an effect that many decades ago had been termed by Rolf Landauer of IBM "information pollution". We have considered one possible mechanism that may curb the unlimited growth, namely the increasing inability of people to convert SHI into PI. This situation may change, however, with an increasing "smartification" of cities including homes, offices, etc. We will return to this issue in Chap. 14.

10.6 Conclusions

'PT and the city', as noted at the outset of this chapter, 'is a very wide issue', referring on the one hand, to grand-scale events such as the first emergence of cities, while on the other, to smaller scale urban phenomena such as suburbanization or gentrification noted above. As we've seen, in the various scales, the PT processes start bottom-up out of an interplay between two kinds of parameters: CP and OP. Our aim in this chapter was to study the intricate relations between the two. We did so descriptively by reference to case studies and formally by two mathematical models: demographically 'growing cities' and urban dynamics as IP. While the mathematization and quantification of the two urban processes gave us a useful tool to explore these complex urban processes, we are fully aware of the existence of other non-quantifiable forces and parameters that play similar roles in PT processes and the city. For example, the planning laws and regulations that stand at the center of the case studies in Sect. 10.2, do not lent themselves to quantification, yet they function as CPs and OPs. The insight from the mathematical models guides us in discussing them qualitatively as will further transpire from Chap. 15 on the planning implications of our study.

A basic element in our synergetic cities view is that the emergence of an urban OP is just 'the first half of the story'. The second half is that once an OP emergences in a city, it not only describes the system city but also prescribes, that is "enslaves", the behavior and action of the parts—the urban agents, and so on in circular causality. While this statement holds quite generally true, it needs further specifications of the kind of city OP we are dealing with. For instance the OP "number of citizens" that we use in our 'growing cities' model (Sect. 10.4), steers/regulates the behavior of citizens responding to housing, working, living conditions, i.e. it enslaves a certain "behavioral pattern". In turn, this OP is brought about by this specific collective behavior. However, there are (many) other behavioral patterns/features of individuals that are influenced by the individuals' interaction with a social group. An explicit example is the feature "urban regulatory focus" that we will deal with in Chap. 11. In this case, the OP "city size" is not sufficient to capture the whole process, but rather we have to introduce some kind of OP hierarchy . Our approach may also shed light on the phenomenon of radicalization. These relations between the slaving principle, circular causality and urban regulatory focus form the content of the next chapter.

References

Bell, G. (1962). *Change in city size distribution in Israel.* Ekistics XIII, 76 p. 103.

Benguigui, L., Czamanski, D., Marinov, M., & Portugali, J (2000). When and where is a city fractal. *Environment Planning B, 27*(4), 507–519.

Bohm, D., & Peat, F. D. (1987). *Science.* Order and Creativity. Bantam, New York.

Florida, R. (2002). *The rise of the creative class: And how it's transforming work, leisure, community and everyday life.* New York: Perseus Book Group.

Hartt, M. (2018). The diversity of North American shrinking cities Urban Studies 2018, Vol. 55(13) 2946–2959.

Jefferson, M. (1939). The law of the primate city. *Geographical Review, 29*(2), 226–232.

Pallagst, K., Wiechmann, T., & Martinez-Fernandez, C. (2014). *Shrinking cities international perspectives and policy implications.* Routledge.

Portugali, J. (1993). *Implicate relations: Society and space in the israeli-palestinian conflict.* Dordrecht: Kluwer Academic.

Portugali, J., & Stolk E. (2014). A SIRN view on design thinking—an urban design perspective. *Environment and Planning B: Planning and Design 2014, 41*, 829–846.

Snow, C. P. (1964). *The two cultures and a second look.* Cambridge: Cambridge Univ Press.

Schwab, K. (2016). *The fourth industrial revolution.* Kindle Edition. Switzerland: World Economic Forum.

Chapter 11
The Slaving Principle, Circular Causality and the City

11.1 Order Parameters and the City

According to Synergetics, the interaction between the parts of a system gives rise to an OP that once emerged, enslaves the parts and so on in circular causality. As we've seen in the case studies described in Chap. 10 above (e.g. balconies, lofts,..), the slaving principle and the associated circular causality took the form of a space-time diffusion process: they started at a certain space–time point, from which they were then diffused throughout the city and other cities (like the corona). What is specifically interesting is, that during this slaving/circular causality process, the OP itself is subject to local variations. For example, if we look at language (e.g. English) as an OP, its space-time diffusion entailed local variations: English–English, American E, Australia E. ... S. African E. ... Or if we look at urbanism as OP, its space-time diffusion entailed local variations: European cities (English, French, ...), American cities, Chinese cities ... they are all similar but with variations.

As noted in Chap. 2, the diffusion form the slaving and circular causality processes take in the dynamics of cities, is a consequence of the property of cities as complex hybrid systems. That is, that cities are composed of artifacts (simple systems) and humans (complex systems). Now, artifacts of all kinds, including buildings, roads and whole cities, are subject to *cultural evolution*—a process that attracted intensive research (Creanza et al. 2017). Of specific relevance here is Cavalli-Sforza and Feldman (1981) study *Cultural Transmission and Evolution: A Quantitative Approach*. Commencing from a neo-Darwinian point of view, they propose that just as biological evolution is governed by *genetic transmission*, so cultural evolution is governed by *cultural transmission*. However, unlike biological transmission, which occurs between parent and child of successive generations only, in the cultural domain transmission takes place between parent and child of successive generations, between neighbors, and between neighbors of two successive generations. These are termed, respectively, *vertical*, *horizontal* and *oblique* transmissions. Similarly to "copying mistakes" (i.e. mutations) in genetic transmission, in culture a new cultural trait is

H. Haken and J. Portugali, *Synergetic Cities: Information, Steady State and Phase Transition*, Springer Series in Synergetics,
https://doi.org/10.1007/978-3-030-63457-5_11

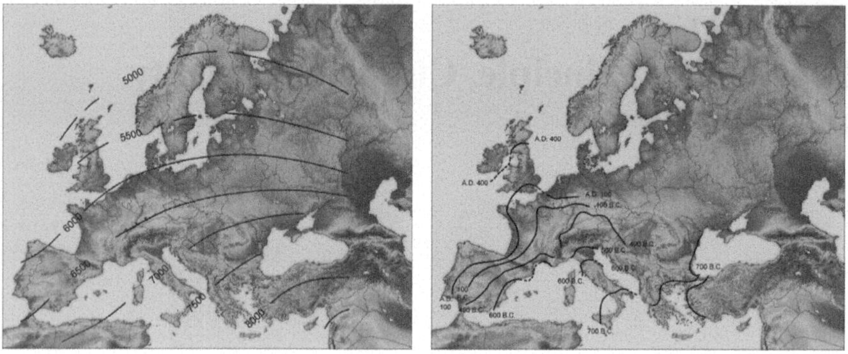

Fig. 11.1 *Left:* the spatial diffusion of agriculture from its core of origin in the midlle east westward. The isolines represent time BP (before present). *Right*: the spatial diffusion of urbanism westward (mainly in the roman period) (After Portugali 2011)

born out of *learning mistakes*. Finally, as a genetic mutation becomes subject to environmental selection, so a newborn cultural mutation becomes subject to selection by the "cultural environment" that determines its fate to extinction or *spatial diffusion*. As an example, Cavalli-Sforza and Feldman reconstructed the spatial diffusion of agriculture Fig. 11.1 left. On Fig. 11.1 right we've added the spatial diffusion of urbanism.

Compared to biological evolution, cultural evolution (that is, the spatial diffusion of cultural traits) is an extremely fast process. There are, however, considerable time differences between such processes. Thus, the spatial diffusion of agriculture and urbanism (Fig. 11.1) are of the time-scale of thousands of years, whereas that of the new ICT (information-communication technologies) gadgets (or the corona) are at the time-scale of months. The speed of adoption depends, on the one hand, on the nature of the cultural trait (e.g. urbanism vs smart phone or coronavirus), while on the other, on the properties and spatial distribution of the potential adaptors. For example, in the case of agriculture and urbanism, the process implied that more and more socio-spatial-cultural communities had to undergo their specific cultural revolution—PT in the language of complexity: from hunting gathering to agriculture in the case of the Neolithic/Agricultural revolution; and from nomadic or sedentary agriculture to cities, in the case of the urban (PT) revolution. The outcome: the space-time diffusion of urban society, for instance, evolved as a sequence of smaller scale PTs, when each such community underwent its own specific urban revolution. For archaeological evidence of such a process see Portugali 2021; the Bedouin community in Israel is currently undergoing an urban revolution, from semi-nomadism to cities (see Lithwick et al. 2004).

To the above we should add, that the spatio-temporal diffusion of innovative cultural entities, evolves in line with our SIRNIA model described in Chap. 4 above: At its core is an interaction between two flows of information: externally represented information that comes from the environment and internally constructed information that comes from the person's and/or social group's memory. Furthermore, in the information adaptation component of the SIRNIA process, each person and each society interprets, and thus adapts to, the innovative cultural trait in a somewhat different way that is affected by its past tradition and experience as constructed and represented in its memory. The result: each society interprets, adapts to, the new innovative trait in its own specific way which in the case of cities entails a variability of urban forms and cities: e.g. European, American, Chinese cities: they are all similar, all dominated by an urban OP, yet they also differ from each other and so on.

Finally it is important to add, that as noted in Chap. 2, the synergetics' processes of slaving and circular causality are similar to what in social theory oriented urban studies is termed social, spatial and socio-spatial *reproduction*. Sources of influence here are, for example, Lefebvre's (1974) *The Production of Space* and Giddens' (1984) theory of *structuration*. Thus for Lefebvre, "biological reproduction and socio-economic production together constituted social reproduction, that is to say, the reproduction of society as it perpetuated itself generation after generation,..". Giddens refers to the interplay between 'structure' and 'agency' which in the context of cities implies the interplay between city structure and urban agents. "Structure", he writes, "is both medium and outcome of reproduction of practices. Structure enters simultaneously into the constitution of the agent and social practices, and 'exists' in the generating moments of this constitution…". The similarity to Synergetics' view is apparent: by means of the reproductive processes of the slaving principle and circular causality, the city's OP describes and prescribes the behavior of the urban agents. This parallelism is significant as it has the potential to bridge the gap between the two cultures of cities noted in Chap. 2 and open the way for CTC to participate in the discourse on the qualitative aspects and problematics of 21st century urbanism.

11.2 Behavioral Features May Cause an OP Hierarchy

While Sect. 11.1 provided the reader with a wide scope of OP diffusion processes over many time- and space-scales, here we turn to the special case of cities. A city is a complicated web of citizens with their various behavioral features and a variety of artifacts (cf. Chap. 2). In the context of our book we are interested in the way this web is coming into existence by self-structurization, i.e. without specific planning. Quite clearly, to characterize the fine-structure of this web an approach based on OPs referring only to gross features of a city such as total number of citizens, total income, productivity etc. is not sufficient. Rather we have to resort to more specific indicators/quantifiers. To bring out our basic idea we consider an explicit example that can be generalized—*cum grano salis*—to a number of other cases including large time- and space-scales.

Such an indicator/quantifier can be derived from the notion *urban regulatory focus* (URF) that we describe and study in some length in Chap. 13 below. As we show there, the notion of URF was originally suggested by Ross and Portugali (2018), then studied and elaborated in the context of synergetic cities by Haken and Portugali

(2019); here we focus on its elements that play a role in the present discussion on slaving and circular causality in a city.

URF commences from Higgin's (1997, 1998) regulatory focus theory (RFT), according to which individuals' goal directed behavior is regulated by two distinct and independently operating, motivational systems—*promotion* and *prevention*. Higgin's RFT further showed that each individual can be characterized by a specific mix of promotion-prevention tendencies termed *chronic regulatory focus*. Some people are thus promotion-oriented, while others, prevention oriented. Promotion oriented individuals are assertive, focus on winning and tend to take risks in order to achieve their goal, whereas prevention oriented individuals, in order to fulfill their aims and goals, prefer to avoid risks and to focus on not making mistakes. While Higgin's RFT refers to individuals only, Faddegon et al. (2008) demonstrated empirically that one's promotion-prevention configuration depends not only on one's chronic regulatory focus, but also on the atmosphere of the group one belongs to. Based on these findings, they have suggested that promotion and prevention can characterize whole groups thus giving rise to a *collective regulatory focus*. Faddegon et al. study referred to small groups such as people in a workplace. Ross and Portugali have taken the issue to the urban domain, showing by means of a set of laboratory experiments, that cities differing in their atmosphere and 'pace of life' affect their citizens' chronic RF. They have measured the effect of a city atmosphere on a person's chronic RF by means of a *response bias*. In their model of urban dynamics, Haken and Portugali (ibid.) have termed the response bias variable '*b*', and employed it as an OP in their urban simulation model.

Now we have all the ingredients together to formulate our general concept by means of this specific RF example. The conventional synergetics' scheme

$$\text{OP} \rightarrow \text{enslaved parts}$$

must be generalized to a hierarchy of OPs that is illustrated graphically in Fig. 11.2. As can be seen in the 3-layers hierarchy of Fig. 11.2, a specific feature (behavior) such as RF, quantified by a bias parameter *b* plays the role of the top OP. The 2nd layer represents two groups of individuals, differing in their specific RF. One group, for

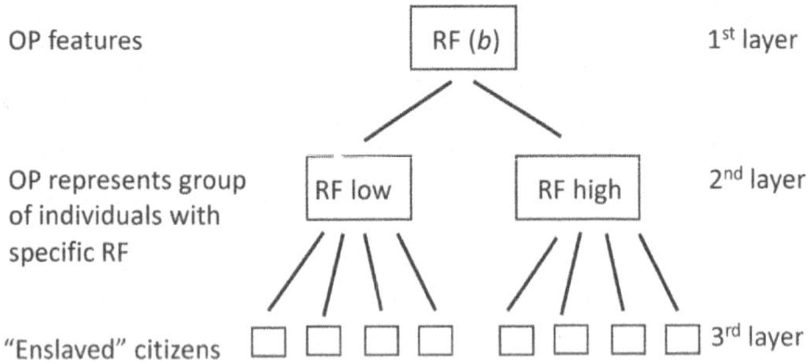

Fig. 11.2 A 3-layer hierarchy of OPs

example, might refer to the citizens of a small city characterized by slow pace of life (low collective RF), while the other to the citizens of a large city that is characterized by fast pace of life (high collective RF).

According to the slaving principle, the behavior of the individuals/citizens of the 3rd layer is enslaved by the OPs of layer 2 that in turn are enslaved by the top OP. Now, according to the circular causality principle, the OP at the top level is fixed by the total behavior of the individuals (3rd layer). As we will detail below, the properties of the ensemble of citizens are represented by the distribution function $P(b)$ of its citizens with bias (features) b. The corresponding ensemble of citizens then determines the behavior of each individual.

Because of the interplay *city (group)—individual*, the behavior represented by b, may change in the course of time. This is represented by an OP equation for b of the form

$$\frac{db}{dt} = -\frac{\partial V(b)}{\partial b} + F(t)$$

where $F(t)$ is a fluctuating force. We will derive $V(b)$ below for our RF-example. As it turns out, V represents an attractor landscape with two valleys, that correspond to a low and high bias value. To each valley, we may attribute an OP according to Fig. 11.2 2nd layer.

According to URF, each specific behavioral pattern is connected with a specific city, i.e. also *different location within a city* (e.g. a quiet suburb vs. very active city center). This allows us to interpret and model the formation of these two cities as a diffusion process in analogy to the diffusion of a particle with position b in a potential $V(b)$ under impact of random "kicks". Mathematically, this process can also be dealt with by studying the time-evolution of the distribution function $P(b)$ (see below). Clearly, this model based on two links/valleys can be generalized to several links that allow the spatial spread of an OP over some distance.

After this summary of essential results of our following RF-approach/model, we discuss generalizations.

1. Employing RFT or URF is but one way to deal with human behavior. Quite clearly, there is a whole set of behavioral patterns/features. In a number of cases, these features are quantifiable. An example is provided by the walking speed of pedestrians (cf. Sect. 6.5). In other cases, we may define scales the way it is done in psychology. RF is just one example. Though the scales may be discrete, we may interpolate between steps so that we may attach a continuous variable to the "size" of the feature. This holds also for physical or mental skills. In all these cases we may derive or at least formulate OP equations (see also below).

If the features are completely or approximately independent of each other, the OPs are more or less independent of each other as well and we can deal with them separately. For each of them, ξ, there is a potential landscape $V(\xi)$. If it possesses only one valley, there is no need to introduce an OP-hierarchy and we may proceed as elsewhere in our book. On the other hand, in case of several (≥ 2) valleys, we have to introduce the new layer.

To each valley we attach its OP that obeys an equation whose potential landscape consists of only one valley (see also below).

It must be emphasized that there is a whole class of behavioral patterns/attributes that don't belong to the just discussed categories. Examples are credo, ideologies, languages, a.s.o.

Finally we mention that diffusion processes can take place at two different levels:

(a) The feature/property in question is transported physically, i.e., by migration of people
(b) By communication only

There are many phenomena that are based on the cooperation or competition of features (see Sect. 11.3).

11.3 Language and the City

11.3.1 OP and Circular Causality

As noted in Chap. 3, the language of a nation has all the typical properties of an OP. It describes a common features of the parts (individuals) of a system (nation/state/city) and it prescribes (enslaves) the behavior of the parts: After birth a baby is subject to the language of his/her parents ("mother tongue"), learns it, and carries it further. The long-living entity language enslaves the behavior of the short-living parts (slaving principle based on time-scale separation). On the other hand, the existence of this very language is brought about by the collective actions of the individuals (circular causality). Note that the learning of this specific language is a prerequisite for the individual's survival.

As further noted in Chap. 3, there are intimate links between languages and cities. In particular theories such as Hillier's (2016) 'space syntax' and Alexander's et al. (1977) 'pattern language', suggest that the morphology of a city is literally a language—a morphological language of cities. Influenced by Chomsky's (1965) *universal grammar*, Alexander further suggested that at the core of all city languages there is a universal urban language—*The Timeless Way of Building* (Alexander 1979). Whether one accepts Chomsky's (and thus Alexander's) view, or not (Tomasello 2003), the implication is that all that has been just said above about language, can be directly applied to cities and to a city's OP. For example, we can rephrase the above sentence as follows: 'After birth a baby is subject to the morphological language of his/her environment, learns it and carries it further'. The environment might be a city, a village or in some areas (e.g. Amazonia) still a jungle.

When a person learns a language, he or she learns (among other things) its syntax, semantics and pragmatics. As we've seen in Chap. 4, a city conveys quantitative syntactic information that can be measured by means of SHI and two forms of qualitative information that we've termed SI and PI. To learn a city is therefore, to learn (among other things) the SHI, SI and PI it conveys. To this one might add, that similarly to a spoken language, and as implied by Aesop's Fable "The Town mouse and the Country Mouse", the learning of this specific city language is a prerequisite for the individual's survival.

The links between language and cities can be further elaborated. Just think of the fact that a city is full of language and texts; each city, neighborhood and street has a name, roads are full of textual instructions (stop, no-entry, …); that in many cities different parts/neighborhoods of the city differ in both their morphological language ("face of the city") and the language spoken in them (e.g. China Town, little Italy). In fact, the dyad city-language requires a separate study that is beyond the scope of the present book.

How can we quantify the OP language? The only way known so far is by means of the number of individuals who speak that language. And how can we quantify the OP city? As we've seen above, one possible way is by means of the number of inhabitants. Under the simplifying assumption that all the members of a nation/society/state and city speak the same language, we immediately can formulate an OP equation for a language and/or a city: it is the Verhulst equation

$$\frac{dq}{dt} = \lambda q - q^2 \tag{11.2}$$

where q—is, in the case of a city, the number of citizens, while in the case of language, the size of the OP "language". λ and γ are control parameters. In case there is an immigration of people who already speak the considered language (e.g. from the rural countryside to big cities), we may add to (11.2) a fluctuation force $F(t)$ in accordance to Sect. 10.4.

To take care of realistic situations (e.g. foreign immigrants arriving to a country's cites), we must widen our approach by taking account of several languages and their interplay. For simplicity, first we consider the process in only one country or even one city. We assume that each inhabitant speaks only one language, say language 1 or 2. Then there are $q_1(q_2)$ individuals who speak language 1(2). Both languages "live" on the same "substrate" total population.

$$q_1 + q_2 = q \tag{11.3}$$

Thus the Verhulst equation (11.2) generalizes to

$$\frac{dq_1}{dt} = \lambda_1 q_1 - \gamma_1 q_1 (q_1 + q_2) \tag{11.4}$$

$$\frac{dq_2}{dt} = \lambda_2 q_2 - \gamma_2 q_2 (q_1 + q_2) \tag{11.5}$$

The coexistence of q_1, q_2 requires

$$q_1 \neq 0, q_2 \neq 0 \tag{11.6}$$

which entails in the steady state

$$\lambda_1 - \gamma_1(q_1 + q_2) = 0 \tag{11.7}$$

$$\lambda_2 - \gamma_2(q_1 + q_2) = 0 \tag{11.8}$$

These equations can only be fulfilled if

$$\lambda_1/\gamma_1 = \lambda_2/\gamma_2 \tag{11.9}$$

which is a very specific condition that implies that the coexistence of languages is a very rare event. A similar phenomenon is known of the coexistence of biological species which live on the same resource. A stable situation is possible only in ecological niches, e.g. by the occupation of different territories. When in a country or a city different languages coexist, then there must be specific "ecological" niches that may be based on more or less pronounced spatial separation, but also on special uses, e.g. scientific language, or dialects used just at home. In these cases, the resources become separated, and (11.4), (11.5) can be replaced by two independent Verhulst equations.

11.3.2 Language Families/Hierarchies

As noted in Sect. 11.1, languages (spoken and urban) can be imported into other countries by immigration which entails a change of language say from English-English to American English etc. While in the new territory, say North America, originally the imported language is the same as that of the country the immigrants stem from, in both parts of the world the languages develop in separated biotopes. Just think of Australia's fauna.

Besides by immigrants, a language (as well as urbanism with specific language of cities) can be brought to another territory also by conquerors, intruders, occupiers who force the subjected people to use the new language either completely or as a second official language. In some cases this adoption of a second official language may happen for practical reasons, for instance as unifying communication means in countries with many languages or dialects. In a number of cases, and at least formally we may define a language (OP) hierarchy such as in Fig. 11.3.

We have said 'formally" because there is a fundamental difference between Figs. 11.2 and 11.3. In Fig. 11.2 we could distinguish between the OPs of the second

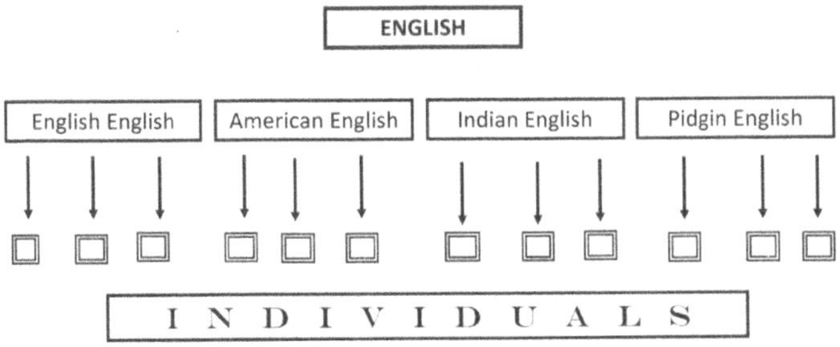

Fig. 11.3 Language's as family resemblance (see below)

layer by means of the numerical value of the bias value "*b*" of RFT which in turn is based on the individuals' behavioral traits, whereas a quantification of the differences between languages is, at least presently, out of reach, and couldn't serve as an explanation anyway. Rather, as noted above the differentiation between different OPs has its roots in the historical development.

Figure 11.3 can also be visualized in terms of Wittgenstein's (1953) *family resemblance category* as a chain of languages/dialects which differ for neighboring territories only weakly, but become at larger distances with more different dialects in between, incomprehensible. Wittgenstein's main concern was 'what makes a category? If you look at the various proceedings that we call "games", he wrote (ibid), "You will not see something that is common to all, but similarities, relationships and a whole series of them at that…". Portugali (2000, 2011) applied the family resemblance notion, in conjunction with the SIRN process, to the evolution and space-time diffusion of the entity/category 'city' in the last 5500 years. The same, as just noted, applies to the space-time diffusion of languages and their dialects. In Fig. 11.3 the OPs of the dialects form layer 2.

11.4 OP Diffusion

As noted in Sect. 11.1, OPs can diffuse from one territory/country to another. One example is language diffusion, another example is the diffusion of urbanism—the language of cities. To derive a diffusion equation for the corresponding OP we recall that a language (and a city) is attached to individuals. We consider the migration of individuals from one territory/city 1, to another one 2. The number of individuals in $j, j = 1, 2$, is denoted by $q_j, j = 1, 2$. Since in the migration process the number of individuals remains constant,

$$q_1 + q_2 = N, N\text{-total number of sindividuals} \tag{11.10}$$

The random migration process can be modelled by means of a *master equation* of the general form (6.15) Sect. 6.1. Here we may specialize the state vector q to q

$= (q_1, q_2)$. Then we are concerned with a master equation for $P(q_1,q_2;t)$. Because of (11.10), we may eliminate q_2, so that we have to deal with a probability distribution $P(q_1; t)$. For a short-hand notation we write q instead of q_1. Note that q is now a *single* variable. The master equation of $P(q;t)$ describes the change of $P(q;t)$ in a very short time-interval. Therefore it is highly improbable that q changes by the simultaneous migration of two or more individuals. In other words, we need to consider changes of q by only one "unit". In accordance with (6.15) Sect. 6.1, we denote the transition probability per unit time from a state q to a state q' by $w(q';q)$. (Note that w must be read from right to left). Having in mind our above remarks on the possible transitions, we arrive at the following master equation

$$\begin{aligned}\dot{P}(q;t) &= w(q;q+1)P(q+1;t) + w(q;q-1)P(q-1;t)\\ &\quad - P(q)(w(q+1;q) + w(q-1;q))\end{aligned} \tag{11.11}$$

We may simplify the notation by the following observation:

$w(q;q+1)$ means that the initial state $q+1$ is lowered by one unit.

Thus we may introduce the notation

$$w(q;q+1) = w_-(q+1)$$

and simultaneously

$$\begin{aligned}w(q;q-1) &= w_+(q-1)\\ w(q+1;q) &= w_+(q)\\ w(q-1;q) &= w_-(q)\end{aligned}$$

With these denotations, (11.11) acquires the form

$$\begin{aligned}\dot{P}(q;t) &= w_-(q+1)P(q+1;t) + w_+(q-1)P(q-1;t)\\ &\quad - (w_+(q) + w_-(q))P(q;t)\end{aligned} \tag{11.12}$$

Now we perform the decisive steps: Since in practice q is a large number, "*1*" can be considered as a small quantity ε. To express this formally in (11.12), we replace there "1" by ε and expand the right hand side of (11.12) into a Taylor series up to 2nd order, where we use

$$P(q \pm \varepsilon) = P(q) \pm P'(q)\varepsilon + \frac{1}{2}P''(q)\varepsilon^2 \tag{11.13}$$

$$w_-(q+\varepsilon) = w_-(q) + w_-'(q)\varepsilon + \frac{1}{2}w_-''(q)\varepsilon^2 \tag{11.14}$$

$$w_+(q-\varepsilon) = w_+(q) - w_+'(q)\varepsilon + \frac{1}{2}w_+''(q)\varepsilon^2 \tag{11.15}$$

The dash' indicates derivative with respect to q. The next steps—inserting (11.13)–(11.15) in (11.12) and rearranging terms—is a bit cumbersome so that we just quote the final result

$$\dot{P}(q;t) = \frac{d}{dq}((w_-(q) - w_+(q))P(q;t) + \frac{1}{2}(w_-(q) + w_+(q))\frac{d^2P(q;t)}{dq^2} \tag{11.16}$$

Because q refers to the number of individuals, $q \geq 0$. This implies $P(q;t) = 0$ for $q < 0$ and the boundary condition $P(0;t) = 0$.

Evidently (11.16) is a Fokker-Planck-$\hat{I}to$ equation where we may identify

$$w_-(q) - w_+(q) = -K(q) \tag{11.17}$$

$$w_-(q) + w_+(q) = Q(q) \tag{11.18}$$

The explicit calculation of w_-, w_+ or equivalently K and Q must be left to special models. One such model was sketched in Sect. 11.2 and will be treated in more detail in Chap. 13. A particularly simple case results if $w_-(q) = w_+(q) = w = const.$

If $K \neq 0$ this may reflect e.g. a social pressure in territory 1 (remember $q \equiv q_1$!) so that people tend to emigrate. But this must be left to more detailed social studies and models.

A final note may be in order: the diffusion processes eventually lead to a steady state where $\dot{P}(q;t) = 0$.

11.5 Concluding Remarks

Cities, as emphasized in Chap. 2 and subsequent chapters, differ from material and organic complex systems in that they are hybrid complex systems. This difference shows itself also with respect to the properties of their OPs, slaving and circular causality in the context of cities: Following its emergence, an OP undergoes space-time diffusion processes that entail variations in its content. While this property was identified by us with respect to cities, it turns out that this is a property of many other systems associated with human culture and society. In this chapter we considered the example of languages in conjunction with the dynamics of cities. Yet it seems that similar considerations apply to other manifestations of collective human culture and behavior (cf. Sect. 11.1) that will have to be explored in the future.

References

Alexander, C. (1979). *The timeless way of building*. New York: Oxford University Press.

Alexander, C., Ishikawa, S., & Silverstein, M. (1977). *A Pattern language*. New York: Oxford University Press.

Cavalli-Sforza, & Feldman (1981). *Cultural transmission and evolution: A quantitative approach*. Princton NJ: Princton Univ Press.

Chomsky, N. (1965). *Aspects of the theory of syntax*. MIT Press.

Creanza, N., Kolodny, O., Feldman, M. W. (2017). Cultural evolutionary theory: How culture evolves and why it matters. *PNAS, 114*(30), 7782–7789.

Faddegon, K., Scheepers, D., & Ellemers, N. (2008). If we have the will, there will be a way: Regulatory focus as a group identity. *European Journal of Society Psychology, 38*, 880–895. https://doi.org/10.1002/ejsp.483.

Giddens, A. (1984). *The constitution of society. Outline of the theory of structuration*. Ploity: Cambridge.

Haken, H., & Portugali, J. (2019). A synergetic perspective on urban scaling, urban regulatory focus and their interrelations. *Royal Society Open Science, 6*, 191087. https://doi.org/10.1098/rsos.191087.

Higgins, E. T. (1997). Beyond pleasure and pain. *American Psychologist, 52*(12), 1280–1300.

Higgins, E. T. (1998). Promotion and prevention: Regulatory focus as a motivational principle. In M. P. Zanna (Ed.), *Advances in experimental social psychology* (Vol. 30, pp. 1–46). New York, NY: Academic Press.

Hillier, B. (2016). The fourth sustainability, creativity: Statistical associations and credible mechanisms. In J. Portugali & E. Stolk (Eds.), *Complexity, cognition, urban planning and design, Springer proceedings in complexity*, pp. 75–92.

Lefebvre, H. (1974/1995). *The Production of Space*. English translation, 1995. Oxford: BlackwelCl.

Lithwick, H., Abu Saad, I., & Abu-Saad, K. (2004). A preliminary evaluation of the Negev Bedouin experience of urbanization: Findings of the urban household survey. Negev Center for Regional Development.

Portugali, J. (2000). *Self-organization and the city*. Berlin/Heidelberg/New York: Springer.

Portugali, J. (2011). *Complexity, cognition and the city*. Berlin/Heidelberg/New York: Springer.

Portugali, J. (2021). The second urban revolution (In preparation).

Ross, G. M., & Portugali, J. (2018). *Urban regulatory focus: A new concept linking city size to human behavior*.

Tomasello, M. (2003). *Constructing a language: A usage-based theory of language acquisition*. Harvard University Press.

Wittgenstein, L. (1953). *Philosophical investigations*. (Translated by Anscombe GEM) Oxford: Blackwell.

Part III
Implications

Chapter 12
Urban Allometry During Steady States and Phase Transitions

Introduction

Allometry—the study of the relations between size, form, physiology and behavior, is recently attracting a lot of attention in the domain of *complexity theories of cities* (CTC). In particular Bettencourt et al. (2007) study attracted a lively discussion that included both criticism (Arcaute et al. 2015; Cottineau et al. 2015, 2017) as well as attempts at elaboration (Batty 2013; Bettencourt 2013; Bettencourt and Lobo 2016; Lobo et al. 2020; Bettencourt et al. 2020). The emphasis in most of the above studies was on socio-economic and infrastructural aspects of urban scaling with little attention to their temporal dimension (e.g. Bristow and Kennedy 2013). Following the study of Depersin and Barthelemy (2017) which focuses empirically on data regarding delay due to traffic congestion and population growth, we see a growing interest in the temporal dimension of scaling (e.g. Keuschnigg 2019). The recent studies of Lobo et al. (2020) and Bettencourt et al. (2020) are attempts to generalize the temporal aspect of urban scaling. Another temporal aspect is the fact that urban scaling studies tend to ignore the evolutionary dynamics of cities between long periods of StS and short events of PT. Elaborating on our study from 2016 (Portugali and Haken, unpublished) our attempt in this chapter is to shed light on the relations between the evolutionary dynamics of cities as complex systems (cf. Chaps. 8–11) and their scaling properties in both steady states and phase transition periods. We do so from the theoretical perspectives of synergetic cities (Chaps. 3, 5–11) and the two associated notions we've developed: SIRN—*synergetic inter-representation network*, IA—*information adaptation* and their conjunction—SIRNIA (Chap. 4).

H. Haken and J. Portugali, *Synergetic Cities: Information, Steady State and Phase Transition*, Springer Series in Synergetics,
https://doi.org/10.1007/978-3-030-63457-5_12

12.1 The Discourse on Urban Allometry

12.1.1 Urban Allometry

Urban allometry commences from the notion that similarly to a natural-organic complex system in which "many of its most fundamental and complex phenomena scale with size in a surprisingly simple fashion" (West and Brown 2005), so is the case also with cities. Namely, "that important demographic, socioeconomic, and behavioral urban indicators are, on average, scaling functions of city size that are quantitatively consistent across different nations and times" (Bettencourt et al. 2007, 7301–2). Based on an extensive body of data in the USA, Germany and China, Bettencourt et al. (ibid.) have proposed, firstly, that most urban indicators can be determined in terms of the following scaling law:

$$Y(t) = Y_0(t)N(t)^{\beta} \tag{12.1}$$

where $Y(t)$ stands for a given urban indicator, $N(t)$ is the population size of a city at time t, and $Y_0(t)$ is a time-dependent normalization constant. Secondly, that the scaling exponent β that characterizes the various urban indicators, can take three universal forms: $\beta < 1$, a *sublinear regime* that typifies economies of scale associated with infrastructure and services (e.g. road surface area); $\beta \approx 1$, a *linear regime* associated with individual human needs (e.g. housing or household electrical consumption); and $\beta > 1$, a *superlinear regime* associated with outcomes from social interactions (e.g. income, number of patents etc.).

12.1.2 The Perspective of Classical Location Theories

The origin of urban scaling can be traced back to the German tradition of classical location theory. In particular to Auerbach's (1913) pioneering study on the rank size distribution of cities and to the more general Zipf's (1949) law, that as shown by Gomez-Lievano et al. (2012) and by Chen (2012), is connected to urban scaling.

Furthermore, from the perspective of the classical location theories there is nothing fundamentally new in Bettencourt's et al. (2007) findings regarding the notions of linear, sublinear, and superlinear regimes: the latter follow from the very logic of these theories. Thus, the linear regime follows from the simple fact that the larger the city, larger is the need for economic activities/products that the inhabitants regularly consume; the sublinear regime from e.g., Weber's (1929/1971) notion of agglomeration and economies of scale; while the superlinear regime from central place theories (Christaller 1933, 1966, Lösch 1954) and the notions of products' *range* and *threshold*. Namely, first, while short-range/threshold economic activities/goods (e.g. bread, milk) will be present in all cities (small and large), long-range/threshold activities/goods (e.g. cars, main government offices) will be present in big cities only. Second, the customers of long-range goods (supplied as they are in large cities only), are not only the registered inhabitants of the large city, but also people that come from nearby, or even faraway, cities and towns. As a consequence, the empirical finding that the number of patents registered in cities scale superlinearly with city size, does not indicate the higher level of creativity of the big city's inhabitants, but rather the simple fact that there is no patent office in smaller towns and cities so that their creative inhabitants register their patents in the bigger cities. The same applies to other 'long range urban activities'.

Note that central place theories use the term "place" and not 'town' or 'city'. Whether or not Christaller or Lösch had the following in mind when choosing the term "place" is an open question, however, the usage of the term *place* is significant for three interrelated reasons: Firstly, it indicates that a city is not a solid entity with clear-cut boundaries, but rather a system and network composed of subsystems. Secondly, that medium- and large-size cities are commonly typified by a hierarchy of central places (Central business district (CBD), neighborhood centers etc.) before being places in a higher level urban system (metropolitan, regional or global). Thirdly, and as a consequence, that the dynamics of a city as a central place, or of central places of a city, is determined not only by its residents, but most importantly, by its users that might come from other (nearby and in some cases far away) cities. Thus in the case of Tel Aviv, not only by its 400.000 residents, but rather by its two million daily users—residents of other towns and cities in the country.

While the origin of urban scaling can be traced back to the classical location theories as above, the current interest in urban scaling differs from the above in that, as noted by Batty (2008), scaling is seen as a mark of cities as complex adaptive systems.

12.1.3 Empirical Reservations and Re-confirmations

The studies that followed Bettencourt et al. (2007) paper, were, and still are, not unequivocal regarding the generality of that study. While some studies support the above findings on urban scaling (e.g. Bettencourt 2013; Bettencourt and Lobo 2016), others have casted doubts about their universality. Approaching the issue from the perspective of complexity, and on the basis of data from France (Arcaute et al. 2015) and England and Wales (Cottineau et al. 2015, 2017), these studies have shown, firstly, "...that population size alone does not provide us enough information to describe or predict the state of a city as previously proposed, ..." and, "that most urban indicators scale linearly with city size, regardless of the definition of the urban boundaries. However, when nonlinear correlations are present, the exponent fluctuates considerably." (Arcaute et al. 2015). Secondly, that "... scaling estimations are subject to large variations, distorting many of the conclusions on which generative models are based." (Cottineau et al. ibid.). In a subsequent study Cottineau et al. (2017) further argue that urban scaling values are not universal, that the socioeconomic diversity within cities is sensitive to size effects and that scaling values vary with city definition. In another study, Depersin and Barthelemy (2018) illuminated the role of past history, namely, that "… strong path dependency prohibits the existence of a simple scaling form valid for all cities".

On the other hand, in a study on Chinese cities, Zünd and Bettencourt (1919) "… show that the distinguishing signs of urban economies—superlinear scaling of agglomeration effects in economic productivity and economies of scale in land use—also characterize Chinese cities", while Sahasranaman and Bettencourt's (2019) study on India's urban system indicates "that many of the results are in line with expectations of the original.." while those who are not are attributed to the uniqueness of India. To the above debate one must add the studies of Lobo et al. (2020) and Bettencourt et al. (2020) that we discuss below.

Most urban scaling studies tend to adopt methodologically inductive, "big data" based, approaches; they commonly commence from data regarding the size distribution of cities and of the various urban indicators. This is so also with the above noted reservations regarding the universality of Bettencourt's et al. (2007) findings—their main concern is the quality of the data. For example, that a major problem is the difficulty (inability?) to accurately delineate the size of cities.

12.1.4 Is the Generality of Urban Scaling by Now Robust?

In two recent studies led by two of the originators of the urban allometry issue (Lobo et al. 2020; Bettencourt et al. 2020), an attempt was made to close the above urban scaling debate. This, by providing empirical evidence accompanied by mathematical formalism that according to the authors prove the generality and robustness of the urban scaling law. The evidence provided in Lobo's et al. (ibid.) study span over thousands of years of urban existence based as they are on archaeological and historical sources as well as on nowadays (big) data. The wide temporal spectrum of this study naturally adds to the discussion to role of time, and thus the second paper by Bettencourt et al. (ibid.) suggests, as the title indicates, an "… interpretation of urban scaling analysis in time".

12.2 Thesis

Regarding the question in 12.1.4, our view is, to rephrase Mark Twain famous saying, that "the reports about the end of the urban scaling debate are greatly exaggerated." Examining the Lobo et al. (ibid.) study we came to the conclusion that their evidence are conceptually, methodologically and empirically problematic, while the mathematical formalism of Bettencourt's el al (ibid.) study leads to contradictions that put the generality of the scaling law in question. In other words, our view is the issue of urban scaling and allometry is still open to a debate.

We further show that the observed exponent fluctuations and the large variations of scaling estimations, are not a result of data deficiencies as it is often assumed: Rather, we demonstrate that scaling variations and exponent fluctuations are a consequence of the very basic properties of the dynamics of cities as complex systems. In other words, we show that *even if the data on city size and urban indicators were perfect, scaling estimations would still be subject to variations due, as noted, to the very basic dynamics of cities as complex systems.*

We demonstrate this, by exploring urban allometry and scaling from the perspective of the "*longue durée*" of cities that is typified by long periods of steady-state (Chap. 8), during which there are small-scale random fluctuations, interrupted be relatively short periods of strong fluctuation that often lead to phase transition (Chaps. 9 and 10). We show, firstly, that while the "classical" scaling law is typical of the steady-state periods of cities, the strongly fluctuating phase transition periods, are typified by other scaling relationships. Secondly, that in cities, such random fluctuations result from the very basic cognitive-behavioral capabilities and tendencies of the urban agents. We demonstrate the first by means of analogies to theoretical and empirical findings in physics regarding the dynamics of material complex systems (Chap. 9). We demonstrate the second, by reference to our above noted SIRNIA approach (Chap. 4), which is a deductive approach that starts theorizing about the city from the bottom–up: from data about humans' innate cognitive and behavioral tendencies.

12.2.1 New Challenges

Why study scaling during periods of strong fluctuations and phase transition?

A major message of urban allometry studies is that a science of cities requires resorting to macroscopic indicators describing gross features as number of inhabitants (city size), total income, total production rates etc. In the spirit of allometry, simple relationships between such indicators are sought for, in particular with reference to city size. Based on large data, allowing for sufficient error margins and using curve fitting, power laws such as (1.1) have been derived empirically. Under the assumption of steady-state conditions, the theoretical derivation by Bettencourt et al. (2007) has been presented. Both the empirical and theoretical urban allometry studies are based on the assumption that in each case the "behavior" of a city is close to the "mean behavior", i.e., that larger fluctuations, deviations from the norm, can be disregarded. However, as current urban reality indicates this is not anymore the case. There are strong indications that we are approaching such an exceptional situation on a large scale.

Just think of the CO_2 problem. There is an increasing public consensus under the motto "we can't go on as always" (an interesting example of the formation of public opinion). Collective decisions must be taken on quite different levels ranging from the city's council's prohibition of diesel cars in cities till those of the world climate conference with still meager results, however. All the required measures require a comparatively fast change of control parameters, such as traffic regulations, taxation on fossil energies such as oil, coal, gas, taxation on energy consumption etc. As we know from our studies in synergetics (see below), fast changes of control parameters entail large deviations from normal behavior (that has been in the focus of "main stream behavior").

As noted in previous chapters, so far, in CTC little or no attention has been paid to such phase transitions that involve—also in urbanism—structural changes, e.g. of the energy supply system, such as gasoline to electricity, traffic etc. They may be accompanied by general economic and social changes, including that of population dynamics (see below). Clearly, there is a need to further develop a theory of phase transitions of cities. In particular we have to discuss the question (alluded to above) in how far the "classical" scaling laws, in particular (12.1) are still valid.

The discussion in the reminder of this chapter thus evolves as follows: The next Sect. 12.3 questions the empirical and mathematical validity and generality of the urban scaling suggested by Lobo et al. (2020) and Bettencourt et al. (2020). Section 12.4 continues Chaps. 9 and 10 by looking at analogies between scaling laws at phase transition periods in physics and in cities. The notion of SIRNIA, that is discussed in length in Chap. 4, provides the theoretical basis for human behavior in cities and the implications thereof to the information conveyed by a city and to the nature of fluctuations in cities. As suggested there, urban dynamics is seen as a process of information production (IP). Section 12.5 adds to the latter a discussion on the implications to the relations between the size distribution of cities and IP. Section 12.6 closes the chapter with a discussion and preliminary conclusions.

12.3 The Reports About the End of the Urban Scaling Debate are Greatly Exaggerated

12.3.1 Cities Form a Family Resemblance Category

A central issue in the debate over urban scaling concerns the definition of the city, its size in terms of area and population—municipal, functional etc. Different definitions of cities might entail different scaling values. The Lobo et al. (ibid.) paper brings another aspect which is more philosophical and cognitive to the fore: what constitutes a city? The opening sentences of Lobo answer as follows:

> There is a general recognition that cities share a number of organisational, social and economic characteristics and play similar functional roles in human societies regardless of size, geography, time or culture. Despite being separated by thousands of years of cultural, social and technological development, settlements of past and contemporary societies seem to share enough in common that the term 'city' can meaningfully refer to both.

This is simply not the case. As shown by Portugali (2000, Chap. 1; 2011, Chap. 10) the various attempts to define cities by means of "shared characteristics" collapsed when put to the Popperian test of *falsification* and eventually the attempts to find such a definition ended up with frustration. This is so with respect to ancient cities, e.g. Childe's (1950) ten-points definition in his seminal paper "The urban revolution", this is so with today's cities, and this is so with the above Lobo's et al. study. How come? The answer comes from discussions on categorization, first in philosophy and subsequently in cognitive science.

The classical view on categories in philosophy that goes back to antiquity (some say to Aristotle) was that categories are 'nomothetic', that is, sets of entities that share some common denominators (necessary and sufficient conditions) that separate/distinguish them from elements belonging to other categories. This view was challenged by Wittgenstein (1953) in his *Philosophical Investigations.* Using the category 'game' as an example, he demonstrated that the elements of the category 'game' are connected not by some common denominators but rather by a network of similarities. As suggested by Portugali (2000, 2011) the same applies to cities. A straightforward way to demonstrate Wittgenstein's view on games and the way it applies to cities, is to use Wittgenstein's own words about the category 'game', but by replacing the word 'games', by 'cities' (in **Bold**):

> Consider for example the entities that we call '**cities**' …. What is common to them all? - Don't say: 'There must be something common, or they would not be called '**cities**', - but look and see whether there is anything common to all. - For if you look at them you will not see something that is common to all, but similarities, relationships and a whole series of them at that. To repeat: don't think but look! - Look for example on [.......]. And we can go through the many other groups of **cities** in the same way; can see how similarities crop up and disappear. And the result of this examination is: we see a complicated network of similarities overlapping and criss-crossing: sometimes overall similarities, sometimes similarities of detail. (Wittgenstein 1953, par. 66)

Wittgenstein termed such categories '*family resemblance*' categories, implying that it is possible that two members of the category have no common properties, yet they are still 'games' or 'cities' by means of the "network of similarities overlapping and criss-crossing".

Wittgenstein's 'family resemblance' provided the starting point to cognitive studies on categorization, to a large extent due to studies by Rosch and coworkers (1976) on categorization and concept formation. Rosch has added that while family resemblance categories are networks as above, they still have a core of *prototypical* elements (e.g. an apple is "more" fruit than a nut). To the above Portugali (ibid.) added that while categories such as games of fruits are studied in a single time period (as they are perceived today), the category city must be looked at in the context of 5500 years of urbanism. Looking at cities from this perspective it has been shown (ibid.) that the core of prototypical cities is changing during time by means of the SIRN dynamics described in Chap. 4, that is, it undergoes phase transitions.

Methodologically, from Popper's view, finding evidence that *verify* one's hypothesis is not sufficient—the task is to prove that there is no evidence that *falsifies* one's hypothesis. In the present case, with respect to Lobo's et al. study, there is plenty of evidence that falsifies the commonality between old and new cities. Here is a short list.

If you look (in Wittgenstein's sense) at ancient 3rd, 2nd and 1st millennium B.C. settlements that are termed 'cities' and then at today's cities, "you will not see something that is common to all ...": Not in their physical structure, size, or socio-economic-cultural properties. Furthermore, as shown by Portugali (1984, 2000, Chap. 15), the very meaning of the word 'city' has been transformed during the years: At the end of the 2nd millennium B.C it was referring to what today we call 'settlement'. There was at that time a distinction between e.g. a 'fortified city', city state, and a 'country city' (e.g. a village in today's language). At a later stage the word 'village' entered and with it a distinction between a city/town and a village and so on. The same with size: a town/city at the end of 2nd millennium B.C. and early 1st millennium B.C. (Portugali 1982, 2000) could be about 10 dunams in size—smaller than the plot size of a single large building in modern cities etc., Furthermore, as noted above, the determination of city size in terms of population is problematic today in the age of "The Data Deluge" (cf. Chap. 2). This is doubly problematic with respect to ancient cities where no population data is available and the only way to get some approximation to population size is to derive it from settlements' size. That is, to speculate on the relations between a city's physical size and population densities (Gophna and Portugali 1987; Portugali and Gophna 1993), The problem is that in most cases exact data on settlement size and population densities is not available. Similarly with the economy—unlike todays cities, ancient cities were essentially based on agriculture—each city had its "fields" (Portugali 1984). The same with social structure, culture and much more.

In light of the above, the generality of the urban scaling claimed by Lobo et al. (ibid.) simply collapses and with it their claim that urban scaling is one among the 'characteristics' that make a settlement city.

12.3.2 How Rigorous are the Scaling Law (1) and Its Derivation?

The law

$$Y(t) = Y_0(t)N(t)^{\beta} \tag{12.1}$$

has been presented in Sect. 12.1 above. It is based on "cross-sectional scaling" that we describe in the following. We consider a set of cities distinguished by a label $i = 1, \ldots N_c$, N_c number of cities. In a first step typical features are measured at *fixed* time t. This procedure is repeated, e.g., every year. In the following t is a fixed parameter. Let Y_i be an extensive indicator that correlates with city size/ population number N_i. We put

$$Y_i(t) = \underset{\{\text{ hypothesis}\}}{Y_0(t)N_i(t)^{\beta(t)}} \; \underset{\{\text{ error correction}\}}{e^{\xi_i(t)}} . \tag{12.2}$$

Though this relation resembles that of e.g. Bettencourt et al. (2020), it differs from theirs decisively. Our indicator $Y_i(t)$ represents its actually measured value, whereas their use of $Y(N_i(t),t)$ instead of Y_i anticipates a specific dependency of Y_i on N_i, which implies a one-to-one correspondance between the label "i" and N_i. This need not be the case: two cities may have the same size whereas their indicators differ greatly. This leads to mathematical inconsistencies (cf. annotations 1 and 3 below).

For any assumed Y_0 or β we always can find a correction factor e^{ξ_i} so that (12.2) is valid for the measured data on Y_i and N_i.We take the logarithm on both sides of (12.2) and put

$$x_i(t) = \ln Y_0(t) + \xi_i(t) \tag{12.3}$$

We obtain

$$\ln Y_i(t) = \beta(t) \ln N_i(t) + x_i(t) \tag{12.4}$$

(Taking the logarithm mitigates the error, cf. annotation 2 below). To determine the parameter β we invoke the standard method of minimizing the mean square error

$$\sum_i x_i^2(t) = min! \tag{12.5}$$

We express x_j by means of (12.4), insert it in (12.5) and seek the minimum by varying β. This leads to

$$\sum_i (\ln Y_i(t) - \beta(t) \ln N_i(t)) \ln N_i = 0 \tag{12.6}$$

or, equivalently

$$\beta = \frac{\sum_i \ln Y_i(t) \ln N_i(t)}{\sum_i (\ln N_i(t))^2} \tag{12.7}$$

A recent publication by Bettencourt et al. (2020) presents a different formula for β that is not derived by means of a variational principle and thus not as rigorous as (12.7). Cf. annotation 3 for a detailed discussion.

Because in our approach there are $N_c + 1$ unknowns, namely $\ln Y_0$ and ξ_i, but there are only N_c equations, we may impose on the unknowns an extra condition. In accordance, e.g. with Bettencourt et al. 2020, we require

$$\sum_i \xi_i(t) = 0 \tag{12.8}$$

So that eventually we obtain

$$\ln Y_0(t) = \frac{1}{N_c} \sum_i (\ln Y_i(t) - \beta \ln N_i(t)) \tag{12.9}$$

where β is given by (12.7). Thus the wanted parameters $Y_0(t)$ and $\beta(t)$ are uniquely determined.

But what about the size and the role of the errors ξ_i? When the scaling law (12.1) is "advertised" it is assumed that the errors ξ_i are small and negligible. First we take this statement for granted and apply it in the following. We first show that the choice of the scaling law may be arbitrary. To this end we observe that our above procedure can be applied when we replace N^β by an arbitrary function $f(N)^\beta$, $f(N)$ e.g., a polynomial of some order with fixed constant coefficients and β a variational parameter. We distinguish different functions by an index k. Then we find quite generally

$$Y(N) = Y_{0k} f_k(N)^{\beta_k} \tag{12.10}$$

(We have dropped the parameter time t).

When we compare (12.10) with (12.1) we obtain

$$Y_0 N^\beta = Y_{0k} f_k(N)^{\beta_k} \tag{12.11}$$

or, equivalently

$${Y_0}^{1/\beta} N = {Y_{0k}}^{1/\beta} f_k(N)^{\beta_k/k} \tag{12.12}$$

This means that any arbitrary function of N can be expressed by a linear function which is a contradiction. It rests on the assumption that errors are negligible, which cannot be the case. In other words, the scaling law holds, at least in general, only up to comparatively large errors. Our example also shows that the choice N^β is rather arbitrary, and other functions $f(N)$ could provide better approximation to $Y(N)$ because f contains more adaptable parameters. All in all our results put the *generality* of the scaling law (12.1) in question.

Annotations

(1) An example is provided by Dutch cities where the number of bicycles per citizens is much higher than in a city of similar size in any other European country. Thus the derivation of a scaling law based on real data crucially depends on a highly biased preselection of cities.

(2) Let X be a random variable with mean X_0 and an error δ,
$X = X_0 + \delta$
where $X_0 \gg 1, \quad \delta \ll X_0$. Then
$\ln(X_0 + \delta) = \ln\left(X_0\left(1 + \frac{\delta}{X_0}\right)\right) \approx \ln X_0 + \frac{\delta}{X_0}$
where $\frac{\delta}{X_0} \ll \delta$ is the new error.

(3) Bettencourt et al. (2020) provide the following formula for β

$$\beta = \frac{\partial ln Y_i}{\partial ln N_i} \tag{12.13}$$

where they put

$$Y_i = Y(N_i(t), t) \tag{12.14}$$

[cf. our discussion after Eq. (12.2)].

The relation (12.13) evidently differs from the best fit (12.7), and is in this form not operational. It implies that changes of Y_i are measured when N_i is changed but such measurements are not done. For an interpretation/evaluation of (12.13) a chain of further hypotheses is required, e.g., similar behavior of different cities (12.14), an assumption still to be proven. Under this assumption, we may arrange city sizes N_i according to their size and establish a map between the set of N and Y. We may try to interpolate so that a continuous plot ln $Y(N)$ versus ln N results. Then we may approximately calculate the slope—but at which value of N? Even if we take the whole (putative) procedure for granted, in the absence of the use of a variational principle, it is arbitrary.

Besides cross-sectional scaling, also temporal scaling has been considered. For a recent paper that establishes a connection between these two kinds of scaling cf. Bettencourt et al. (2020). To establish temporal scaling, for each city i the temporal evolution of Y_i is studied as function of the temporal evolution of N_i. More precisely, this approach posits a scaling law of the form

$$Y(N_i(t), t) = A_i N_i(t)^{a_i} e^{\chi_i(t)} \tag{12.15}$$

where it is postulated that A_i and a_i are time-independent. Again the inclusion of the variations (errors) χ_i renders (12.15) an exact relations. Of course (12.15) is of practical use only if the errors are small.

12.3.3 PTs and the Scaling Laws (12.1) and (12.15)

The "developers" of both cross-sectional and temporal scaling laws, (12.1) and (12.15) ignore the distinction between steady states (StS) and phase transitions (PT), whereas this is a major issue in our book. So to shed light on possible effects of PTs on the laws (12.1) and (12.15), we take them as granted, including the replacement (12.14). We consider several cases of PTs. They may happen in a single city (case

1) or, at the same time, in the majority of cities (case 2). Furthermore, a specific PT is characterized by

(a) N_i is practically constant, whereas the measured Y_i changes strongly within some time-interval.
(b) N_i changes strongly, whereas Y_i is fixed.
(c) Both N_i and Y_i change strongly in the course of time.

Let us discuss these cases in more detail.

We choose $i = 1$. N_1 is (practically) constant, while Y_1 changes dramatically from $t = t_0$ till $t = t_f$. Thus

$$N_1(t_f) = N_1(t_0) \tag{12.16}$$

Y_1 (measured at t_0) $\gg Y_1$ (measured at t_f), whereas the scaling law (12.15) predicts

$$Y(N_1(t), t_f) = A_1(t_0)^{a_1} \tag{12.17}$$

i.e. an unchanged indicator value so that a large error $\chi_i(t_f)$ results and the law (12.15) fails.
We consider the cross-sectional law (12.1). At t_f and with (12.16) we obtain

$$Y(N_1(t_0), t_f) = Y_0(t_f) N_1(t_0)^{\beta(t_f)} \tag{12.18}$$

And for $i \neq 1$

$$Y(N_i(t_f), t_f) = Y_0(t_f) N_i(t_f)^{\beta(t_f)} \tag{12.19}$$

The parameters $Y_0(t_f)$, $\beta(t_f)$ on the r.h.s. of (12.18) remain practically unchanged because they are determined by the majority of cities (12.19). Thus these parameters don't allow the detection of PT. However (12.18) discloses a discrepancy because
$Y_1(measured) \gg Y(N_1(t_0))$ theory.
An analogous consideration holds when for a specific city, say $i = 1$, Y_1 is practically constant, whereas N_1 changes strongly—a case we treat in Sect. 10.4.
What happens when both Y_1 and N_1 grow (or decay) simultaneously so that

$$Y_1(t_f) = \gamma Y_1(t_0) \tag{12.20}$$

$$N_1(t_f) = \delta N_1(t_0) \tag{12.21}$$

The law (12.15) is only fulfilled if

$$\gamma = \delta^{a_1} \tag{12.22}$$

In (12.1) the parameters Y_0 and β remain approximately unaffected.

(2) All cities are involved and the errors ξ_i in (12.2) can be neglected which implies that Y_i must be of the form (12.14). We assume that for the measured values

$$Y_i(t_f) = \gamma_i Y_i(t_0) \tag{12.23}$$

$$N_i(t_f) = \delta_i N_i(t_0) \tag{12.24}$$

We insert (12.23), (12.24) in (12.2) for t_f and use (12.2) again for t_0. We obtain

$$\ln\gamma_i + \ln Y_0(t_0) + \beta(t_0)\ln N_i(t_0) = \ln Y_0(t_f) + \beta(t_f)\ln\delta_i + \beta(t_f)\ln N_i(t_0) \tag{12.25}$$

Because (12.25) must hold for any $N_i(t_0)$, we obtain

$$\beta(t_f) = \beta(t_0) \tag{12.26}$$

$$\gamma_i \delta_i{}^{-\beta} = Y_0(t_f)/Y_0(t_0) \tag{12.27}$$

which is a very stringent condition on γ_i and δ_i. Strong violations are indicators for a collective PT, or of a failure of the scaling law (12.1) altogether.

12.3.4 Conclusion

As it transpires from our analysis in Sect. 12.3.2, the Bettencourt et al. (2020) phenomenological approach rests on the assumption of self-consistency, whereas our approach (cf. 12.5) rests on a variational principle. The former approach may be vulnerable because of counter examples, cf. annotation (12.1) in Sect. 12.3.2.

Our discussion leaves aside attempts at deriving scaling laws based on "microscopic" urban processes (Bettencourt et al. 2020; Lobo et al. 2020), because in our view their assumptions are at least partly, unjustified (cf. our discussion in the beginning of Chap. 12).

12.4 Scaling Laws at Steady States and Phase Transitions

While the derivation of scaling laws has been based on big data for the "normal" state, such data are sparse in case of phase transitions of cities. Thus there is only a shaky basis for any empirical "curve fitting" and we must develop appropriate models. In view of the complexity of urban processes it is advisable to begin with well-studied model systems. These are provided by physical systems with their pronounced phase transitions, such as melting of ice to water, onset of ferromagnetism etc. These transitions occur when e.g. the temperature is changed and show up at the macroscopic level in form of qualitative changes of a number of physical properties.

A prototypical example is provided by a ferromagnet (cf. Chap. 9). At room temperature it possesses a spontaneous magnetization M (with its "north and south poles"). When heated above a critical temperature Tc (called Curie temperature), the magnetization vanishes entirely. The transition curve has been measured very carefully, and empirically a *scaling law* was found (T: temperature)

$$M = const.(T_c - T)^{\beta} \tag{12.28}$$

with a "critical exponent"β.

The derivation of (12.28) including the correct value of β based on a *microscopic* theory turned out to be a major challenge for theoretical physics.

As we show in some details in Chap. 9, on the microscopic level, a ferromagnet is composed of tiny elementary magnets that we denote as "spins" and that can point in only two directions, up and down. It is energetically favorable if two neighboring spins point into the same direction. This effect is counteracted by *thermal* motion. At low temperature this disturbance is small so that all spins are lined up: maximum magnetization. Quite generally, the magnetization is just the difference between the number of "up" and "down" spins, respectively (up to a constant factor). With increasing temperature, more and more spins are flipped and the magnetization decreases. To visualize the emerging spin configurations think of a two dimensional grid of spins we are looking at from above and where up-spins are marked by a black dot and down-spins by a white dot. As computer simulations reveal, close to the transition temperature, the black/white dots represent patterns strongly reminiscent of fractals. This observation provided the key to derive the law (9.28) "from first principles" (cf. Chap. 9).

Fractals can be constructed stepwise by using finer and finer scales, or to put it differently, we may consider fractals with different degrees of resolution. In a ferromagnet a given resolution entails a specific kind of patterns (configurations) of spins. Now, the free energy[1] of ferromagnet must not depend on the degree of resolution (under which we model the magnet). On the other hand, as detailed study reveals, different spin-patterns are connected with different temperatures and magnetizations.

---

[1] The free energy is the difference between the total energy and the energy due to the irregular spin flips ("thermal agitation") (Landau and Lifshitz 1980).

How can, nevertheless, the free energy be conserved? This is achieved by a *coordinated* change of temperature and magnetization when the resolution is increased or decreased. The nature of the transformations leading to fractals (in technical terms a renormalization group) necessitates a power law between magnetization and temperature, i.e. (9.1). This kind of scaling is believed to be of universal nature for phase-transitions for systems in thermal equilibrium.

In how far can we utilize these results when dealing with the state and development of cities? First of all, at least larger cities have a fractal structures, e.g. the networks of streets, water, gas, electricity supply, telephones etc. but the range of the "degrees of resolutions" is rather limited. Furthermore, the laws that connect the different levels are different from a ferromagnet which is in thermal equilibrium. The transportation networks just mentioned operate quite differently. Nevertheless, they may be ruled by scaling laws based, e.g. on principles such as conservation of mass (of all sorts: pedestrians, vehicles, electricity, gas).[2] But presently the basic assumptions must be changed. To mention two examples: WLAN etc. is replacing telephone networks, and three-dimensional traffic (e.g. drones) is more and more established. This implies that it will become more and more difficult to build on past experience.

Returning to physical model systems: there is a second class that exhibits a phase transition far from (thermal) equilibrium. A prototypical example is the light source laser. Its operation is maintained by an influx of energy. At a sufficiently high influx ("threshold") the disordered light of a conventional lamp becomes highly ordered, in other words, the statistical properties of light change dramatically. Below the threshold, the statistics of the light field can be described by a Gaussian with given mean and variance. When, with enhanced energy influx, the threshold is approached, the variance becomes very large: "critical fluctuations". At and above threshold the appropriate probability distribution is no more Gaussian, but requires still higher order powers of the field in the exponent of the former Gaussian. As a consequence, close to threshold, simple scaling laws such as (12.1) can no more be derived. This becomes possible only when we treat limiting cases (cf. Chap. 10).

The laser phase-transition may be a viable model for transitions of city structure/function, because, like the laser, cities are open systems that exchange energy, matter and/or information with their surroundings. Section 10.4 presented an explicit example of a "city transition".

As we've seen, fluctuations are present during both steady state and phase transitions while in the latter they often become "critical" and lead to structural changes. In natural (material and organic) complex systems fluctuations can be likened to random events—a kind of"mutations". Yet in the case of cities, their temporal appearance might indeed be random, but their cause is not a copying mistake, as in organic systems; rather it is a result of urban agents' cognitive capabilities in their ongoing interaction with the city—a process captured by the theory of SIRNIA.

---

[2]Such "items" produced (or enter into the network) at specific rates at a source (producer), are distributed by a fractal network, and eventually consumed (or leave the network) at "sinks" (consumers). Clearly at each bifurcation (branch) point of the network, the "material" flux decreases by a factor ½ in each branch, implying a scaling law.

12.5 SIRNIA and the City

The notions of SIRN, IA and their conjunction SIRNIA were introduced in Chap. 4 as well as the implications of the SIRNIA perspective to urban dynamics. In particular it was shown that from the perspective of SIRNIA, the fluctuations during the steady state periods of cities might often be random, but also a result of urban agents' cognitive capabilities in their ongoing interaction with the city. For example, when looking at a ground floor of a city building, officially defined (e.g. by its architecture and/or by the planning law) as residential, one urban agent might see an apartment—i.e. the PI of living, while a second urban agent sees also the potential to transform this apartment into an office, studio and/or a shop. Our first agent's action thus conforms with the city's plan/order parameter, while that of the second agent does not—it violates the city's order parameter /planning law. Our basic thesis is that such actions imply fluctuations or "urban mutations": small local deviations from the rule (the 'normal") that may lead to (i.e. trigger) macroscopic transformations. When these fluctuations occur during the long steady state periods, their effect is negligible and the "classical" scaling relations prevail; when in the unstable period, (that often leads to phase transition) we observe variations in scaling estimation. (cf. Sect. 10.4).

It was further shown that from the perspective of SIRNIA, urban dynamics is seen as a process of information production (IP)—a process that often gives rise to the city's order parameter. To the latter we add here some further noted on city size and information production.

12.5.1 *City Size and Information Production*

From the perspective of the notion of IP follows that city size (in population) can be related to information in the following ways. First, generally speaking, with the exception of university towns whose function is to produce information (e.g. Cambridge UK or Berkeley US), the larger a city is, larger is its SI and PI, that is, more meaningful aspects and activities it affords, higher is the SI and PI derived SHI it conveys to its own population, as well as to population *outside the city* in its country and in the world. Examples can be found in all aspects of urban life, ranging from the socio-cultural, through the architectural to the economic composition of cities. Thus, as shown in past studies (Portugali 2000, Chaps. 7 and 8), the larger a city is, larger also is the number and variety of the socio-cultural groups that emerge out of its dynamics and make its spatially segregated landscape. In most cases, the larger cities are also the older, with the longer process of urban development, architecture and history. As a consequence, the SI "face" of such cities (i.e. their urban landscape) is relatively more diverse in terms of buildings' and roads' form, size and architecture. Note that the situation where all buildings are architecturally similar (low SI determined SHI as discussed in Chap. 4, Sect. 4.2.2), is typical to new cities, while that of high diversity of buildings form and style (high SHI), to relatively old cities.

Furthermore, as noted above (Sect. 12.1.3), from the principles of classical central place theories, as well as from empirical studies, and the discussion above follows that the bigger the city bigger also is the diversity of its economic activities and thus its PI determined SHI. Generally speaking, this view is in accord with Bettencourt's et al. view (ibid.): the extent to which this is related to Bettencourt's et al. linear, superlinear and sublinear forms, has yet to be studied empirically.

Second, the larger a city, denser is its network of internal and external connections and potential interactions possibilities between its own populations as well as with population *outside the city,* in its country and in the world. As a consequence, more attractive the city becomes to immigrants from its country and from the world, more immigrants will come and reside in the city, faster will be its demographic growth and faster the city will reach its spatial constraint, i.e. its size limit or other limitations. The Barabási–Albert (1999) model of *preferential attachment,* with its *the rich get richer* aphorism, applies here too, that is, that the bigger the city the faster it demographically grows. It is not a surprise that global labor migration is attracted almost exclusively by the economic opportunities (PI determined SHI) conveyed by the biggest cities of European countries (e.g. Rotterdam, the second-largest city in the Netherlands, is home to many ethnic-minority groups, which make up almost half the population of the city, compared to about 15% in a small city like Delft).

The result: urban growth takes the classical form of Verhulst's (1838) logistic curve. Such a growth process might lead to stagnation, unless there is a new incentive for further growth, e.g. new land available, new traffic connections etc., or even the formation of a new opinion on the socio-economic structure of a city, e.g. its use of buildings, reconstruction etc. At this stage the role of fluctuations becomes significant.

12.6 Preliminary Conclusions

In steady state a city is governed by many regularities, e.g. the routinized behavior of its citizens, and it is possible to derive—within certain limits—scaling laws (12.1) for (macroscopic) indicators. However, there are rare cases where strong fluctuations entail, at least partly, a transition to new kinds of behavior (and structures).

In this chapter we scrutinized the question in how far scaling laws also apply to indicators of a city that undergoes a (phase) transition. We first discussed phase transitions of physical systems because of their relative simplicity. In the ferromagnet fractal structures of spins are formed that accompany the transition from a fully ordered magnetic state to a disordered state when temperature is increased. This process might serve as model for the rearrangement of supply networks in cities in response to new requirements.

In the (conventional) laser the spatial arrangement of its atoms, is irrelevant. Here we are interested in the transition from the highly uncorrelated emission acts by the atoms of a lamp to the highly ordered laser light. As theory shows, this is achieved by a strong self-organized coordination of the atomic emission acts. When we apply

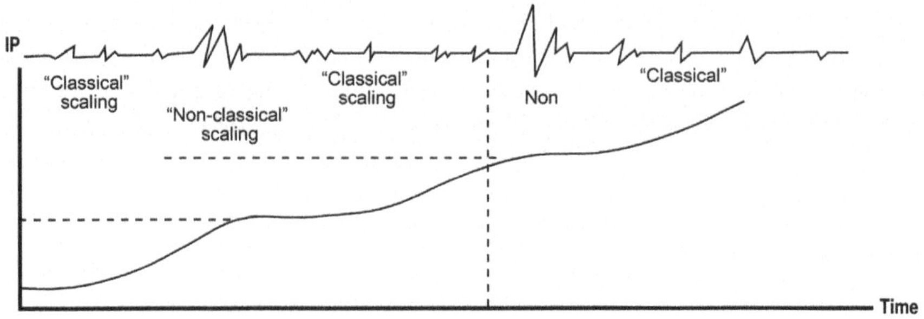

Fig. 12.1 The "classical" scaling law characterizes the city during its steady state; once the city becomes unstable, local fluctuations may give rise to global phase transition and "non-classical" scaling relations

this insight to urbanism, we recognize the fundamental role of the SIRNIA process on the collective level. Future theoretical work will have to illuminate the interplay between temporal, spatial and functional reorganization processes, and to study in how far scaling laws can be derived. In Sect. 10.4 we showed how a new kind of scaling law results when the impact of random events is taken into account.

In conclusion, looking at the issue of urban scaling laws from the conjunctive perspectives of SIRNIA (Fig. 12.1), we can thus see that the "classical" scaling law characterizes the system city during its steady state—in this period (large λ) local fluctuations do not affect/endanger the global structure of the city. The city at this state is resilient against fluctuations. However, when λ is small, that is, when the global structure of the city becomes unstable for reasons described above, then local fluctuations have major effect on the system and may cause/give rise to global change/phase transition .

References

Arcaute, E., Hatna, E., Ferguson, Y. H., Johansson, A., & Batty, M. (2015). Constructing cities, deconstructing scalig laws. *Journal of the Royal Society Interface, 12,* 20140745.

Auerbach, F. (1913). Das Gesetz der Bevoelkerungskonzentration. *Petermanns Geograph Mittl, 59,* 74–76.

Batty, M. (2008). The size, scale and shape of cities. *Science, 319,* 769–771.

Batty, M. (2013). *The new science of cities*. MIT Press Cambridge Mass.

Bettencourt, L. M. A., & Lobo, J. (2016). Urban scaling in Europe. *Journal of Royal Society Interface, 13*, 20160005. https://doi.org/10.1098/rsif.2016.0005.

Bettencourt, L. M. A., Yang, V. C., Lobo, J., Kempes, C. P., Rybski, D., & Hamilton, M. J. (2020). The interpretation of urban scaling analysis in time. *Journal of the Royal Society, Interface, 17,* 20190846. https://doi.org/10.1098/rsif.2019.0846.

Bettencourt, L. M. A., Lobo, J., Helbing, D., Kühnert, C., & West, G. B. (2007). Growth, innovation, scaling, and the pace of life in cities. *Proceedings of National Academy Science U.S.A., 104*(17), 7301–7306. https://doi.org/10.1073/pnas.0610172104.

Bettencourt, L. M. A. (2013). The origins of scaling in cities. *Science, 340,* 1438–1441.

Bristow, D. N., & Kennedy, C. A. (2013). The energy for growing and maintaining cities. *Ambio, 42,* 41–51. https://doi.org/10.1007/s13280-012-0350-x.

Chen, Y. (2012). The mathematical relationship between Zipf's law and the hierarchical scaling law. *Physica A: Statistical Mechanics and Its Application, 391*(11), 3285–3299.

Childe, V. G. (1950). The urban revolution. *Town Planning Review, 21,* 3–17.

Christaller, W. (1933/1966). *Central places in Southern Germany.* Englewood Cliffs, NJ: Prentice Hall.

Cottineau, C., Hatna, E., Arcaute, E., & Batty. M. (2015). Paradoxical interpretations of urban scaling laws. July 2015 ArXiv.

Cottineau, C., Hatna, E., Arcaute, E., & Batty, M. (2017). Diverse cities or the systematic paradox of Urban Scaling Laws. *Computers, Environment and Urban Systems, 63,* 80–94.

Depersin, J., & Barthelemy, M. (2018). From global scaling to the dynamics of individual cities *PNAS, 115*(10), 2317–2322.

Gomez-Lievano, A., Youn, H., & Bettencourt, L. M. A. (2012). The statistics of urban scaling and their connection to Zipf's Law. *PLoS ONE, 7*(7), e40393.

Gophna, R., & Portugali, J. (1987). Settlement and demographic processes in Israel's coastal plain from the Chalcolithic to Middle Bronze age. *Bulletin of the American Schools of Oriental Research, 269*(1987), 11–28.

Keuschnigg, M. (2019). Scaling trajectories of cities. *Proceedings of National Academy Science, 116*, 201906258.

Landau, L. D., & Lifshitz, E. M. (1980). *Statistical physics* (3rd ed., Vol. 5). Butterworth-Heinemann. ISBN 978-0-7506-3372-7.

Lösch, A. (1954). *The economics of location.* New Haven: Yale Univ Press.

Lobo, J., Bettencourt, L. M. A., Smith, M. E., & Ortman, S. (2020). Settlement scaling theory: Bridging the study of ancient and contemporary urban systems. *Urban Studies, 57*(4), 731–747.

Portugali, J. (2000). *Self-organization and the city.* Berlin/Heidelberg/New York: Springer.

Portugali, J. (1982). A field methodology for regional archaeology: The West Jezreel Valley survey 1981. *Tel Aviv, 31*(2), 170–190.

Portugali, J. (1984). Arim, Banot Migrashim Vehazerim: The spatial organization of Israel at the 10–12 centuries B.C., *Eretz Israel, 17*, 282–290.

Portugali, J. (2011). *Complexity, cognition and the city.* Berlin/Heidelberg/New York: Springer.

Portugali, J., & Gophna, R. (1993). Crisis, progress and urbanization: The transition from early bronze I period to early bronze II period in palestine (Late 4th millennium B.C.). *Tel Aviv, 20*(2), 164–186.

Rosch, E., Mervis, C., Gray, W., Johnson, D., & Boyes-Braem, P. (1976). Basic objects in natural categories. *Cognitive Psychology, 8*, 382–439.

Sahasranaman, A., & Bettencourt, L. M. A. (2019). Urban geography and scaling of contemporary Indian cities. *Journal of the Royal Society, Interface, 16,* 20180758. https://doi.org/10.1098/rsif.2018.0758.

Verhulst, P. F. (1838). Notice sur la loi que la population suit dans son accroissement. *Correspondence Mathematical and Physical Publication Par A Quetelet, X,* 113–121.

Weber, A. (1929/1971). *Theory of the location of industries.* Chicago: Chicago Univ Press. (Translated by CJ Friedrich).

West, J. H., & Brown, G. B. (2005). The origin of allometric scaling laws in biology from genomes to ecosystems: Towards a quantitative unifying theory of biological structure and organization. *Journal of Experimental Biology, 208,* 1575–1592. https://doi.org/10.1242/jeb.01589.

Wittgenstein, L. (1953). *Philosophical investigations* (Translated by Anscombe GEM). Oxford: Blackwell.

Zipf, G. K. (1949). *Human behavior and the principle of least effort.* Cambridge MA: Addison-Wesley.

Chapter 13
Urban Scaling, Urban Regulatory Focus and Their Interrelations

13.1 Introduction

13.1.1 The Pace of Life in Cities

The previous chapter on urban allometry started with Bettencourt's et al. (2007) paper, "Growth, innovation, scaling and the pace of life in cities". As the title implies, Bettencourt's et al. study shows that cities differ in their "pace of life", namely, that there are fast- and slow-paced cities—a view that goes back to the classical urban theories of Simmel (1964), Wirth (1938), Milgram (1970) and others. Bettencourt's et al. allometric approach attempted to quantify this view, relate it to city size, and show statistically that (contrary to organic systems) "*the pace of urban life is predicted to increase with [city] size*". As they demonstrated, this property shows itself in the various urban indicators that scale superlinearly ($\beta > 1$) with city size, ranging from innovation and wealth creation to crime rates, rates of spread of infectious diseases such as AIDS, and even pedestrian walking speed.

In a subsequent study, Bettencourt (2013) rederived the scaling laws in a direct way. Based on a set of four assumptions regarding accessible resources to citizens, infrastructure, human mentality and social interaction, he predicted "scaling exponents for a wide variety of urban indicators, from patterns of human behavior and properties of infrastructure to the price of land" (ibid., 6, 1439). By this, he suggested, his modelling approach "ties together the most microscopic needs and behaviors of individuals anywhere to the most macroscopic aspects of the urban infrastructure" (ibid., S19).

When scrutinizing Bettencourt's recent studies we arrived at the conclusions that they rest on (at least) two preconditions: cities are self-similar fractal structures, and, they are subject to laws of large numbers; the implication: the approach applies to large enough cities, but not to small city/town sizes.

H. Haken and J. Portugali, *Synergetic Cities: Information, Steady State and Phase Transition*, Springer Series in Synergetics,
https://doi.org/10.1007/978-3-030-63457-5_13

An alternative interpretation of the relations between the macroscopic properties of urban scaling and the microscopic aspects of urban agents' behavior was recently suggested by Ross and Portugali (2018). Their paper explains 'why' and 'how' the size of cities has an effect on their inhabitants' and users' behavior. Commencing bottom-up from humans' basic cognitive capabilities, they answer the 'why' and 'how' by extending the principles of Higgins' (1997) *regulatory focus theory* (RFT) regarding humans' motivational system, to the context of cities. RFT and its extension to cities as *Urban Regulatory Focu*s (URF) were already shortly introduced in Chap. 11. In what follows we introduce them in more details.

13.1.2 Regulatory Focus Theory and Collective Regulatory Focus

Higgins' *regulatory focus theory* (RFT) regarding humans' motivational system demonstrates that individuals' goal-directed behavior is regulated by two distinct, and independently operating, motivational systems—*promotion* and *prevention* (Higgins 1997). It further shows that while every person is driven by both promotion and prevention, promotion oriented individuals are assertive, focus on winning and tend to take risks in order to achieve their goals, whereas prevention oriented individuals are non-assertive, tend to avoid risks and to focus on not losing.

Higgins' RFT with its promotion-prevention tendencies refers to individuals' basic personal character and it is thus also termed *chronic regulatory focus*. Subsequent studies demonstrated that a person's chronic regulatory focus is context dependent, namely, that groups have an effect on the regulatory focus strategies of their members. In particular, Faddegon et al. (2008) demonstrated empirically that the likelihood of one's behaving in a promotion or prevention way depends not only on one's personal (chronic) regulatory focus, but also on the atmosphere of the group one belongs to. That is, an assertive high-tech company (e.g. Apple company with its "think different" slogan) and a conservative insurance company (whose possible slogan might be "better safe than sorry") affect differently the personal regulatory focus of their employees. Based on these findings, Faddegon et al. (ibid.) have suggested that promotion and prevention can characterize whole groups thus giving rise to a *collective regulatory focus*. This effect was found in the context of small groups where face-to-face interaction prevails as well as in the case of larger groups where face-to-face interaction is rare.

13.1.3 Urban Regulatory Focus

Inspired by the above research, Ross and Portugali (2018) theorized that regulatory focus processes occur at the urban level and suggested a link between RFT and the statistical findings of urban allometry: As noted above, according to urban allometry studies (Bettencourt et al 2007; Bettencourt 2013), large cities are *usually* characterized by fast-paced and competitive urban dynamics, compared to small cities which are *usually* relaxed and slow-paced with a focus on safety, stability, and security.

To test the above theorization, Ross and Portugali (2018) have conducted a set of laboratory experiments in which each subject was, firstly, tested for his/her chronic regulatory focus (by means of Higgins' et al. (2001) questionnaire); secondly, was shown an (image of) urban scene of a fast-paced or a low-paced city (as in Fig. 13.1 top and bottom); and thirdly, was once again tested for his/her regulatory focus. This second test was implemented by means of the *recognition memory task* (a variation of a 'signal detection task'), which together with the *response bias* measure has been successfully used in the past in the context of RFT studies (Faddegon et al. 2008; Crowe and Higgins 1997).

In the present set of experiments, the dependent variable was the participants' *response bias* that measured *the extent to which fast-paced cities versus small-paced cities affected the personal (chronic) regulatory focus of the subjects.* In the model we develop below (Sect. 13.3), we denote the response bias as b. As noted above, while every person is driven by both promotion and prevention, some people are promotion oriented while others prevention oriented. Higher values of the response bias measure (e.g. higher b) indicate an increase in the personal promotion oriented component due to the urban effect, while lower values (lower b), a decrease of the promotion component and an increase of the prevention component, due to the urban effect.

Using the above methodological apparatus in their series of laboratory experiments, Ross and Portugali (2018) have demonstrated that the promotion urban context of fast-paced cities (which are *usually* large cities) tend to intensify both promotion-

Fig. 13.1 Top, scenes from fast-paced cities. Bottom, scenes from slow-paced cities. (Source: Ross 2015)

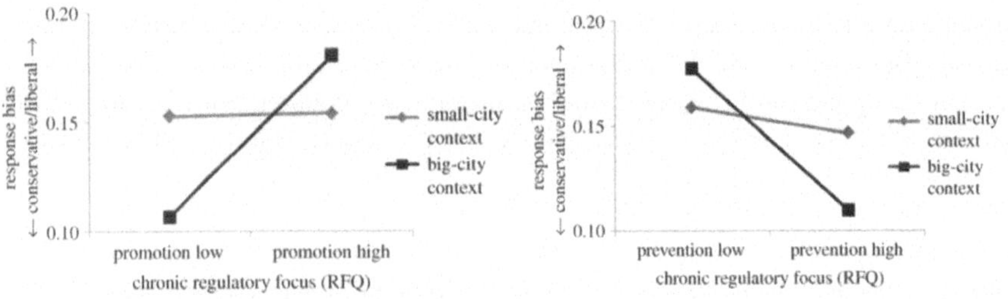

Fig. 13.2 Fast-paced city context intensified promotion focused behavior (left) as well as prevention focused behavior (Right). (Source: Ross 2015; see also Ross and Portugali 2018)

and prevention-focused behaviors, thus motivating individuals to behave in extreme and polarized ways (Fig. 13.2), whereas slow-paced cities (which are *usually* the small-size urban environments), tend to encourage relatively more moderate and less polarized behavior.

Note that we use and emphasize the expression "*usually*" in the above paragraphs. This is to indicate that there are exceptions—small but fast-paced cities (e.g. Oxford, Silicon valley, …). Urban scaling studies such as Bettencout's et al. (2007), refer to generic cases and general statistical regularities, to which there are always exceptions. Ross and Portugali (2018), and thus our present study, refer essentially to slow vs. fast paced cities irrespective of their size, when the link to city size is due to Bettencourt et al. (2007).

13.1.4 Aims

The situation so far is as follows: from urban scaling studies (cf. Chap. 12) we learn that urban allometry with its linear, superlinear and sublinear statistical regularities is a basic characteristic of cities as complex adaptive systems. From Bettencourt's (2013) subsequent study, as well as from other studies (e.g. Chen 2021; Ribeiro et al. 2017), we further learn that the above statistical regularities can be explicitly derived out of the property that cities as complex systems are self-similar fractal structures. This rederivation, however, is implicitly pre-conditioned by the laws of large numbers with the implication that the solution excludes the very small cities and towns in the urban system. Finally, from Ross and Portugali (2018) study about URF we learn that fast- and slow-paced cities (which are usually large and small cities) affect the prevention-promotion tendencies of their inhabitants and users in different ways as above. What still remains an open question following these studies, however, is the way (or the extent to which) these motivational-behavioral reactions are related to the dynamics of cities as complex, adaptive, self-organization systems. How the dynamic of cities of different paces of life (sizes) is affected by, and is affecting, the promotion and prevention tendencies of their inhabitants and users?

Our aim in this chapter is to 'close the circle' and answer this open question. We do so from the theoretical perspectives developed in the previous chapters, that is, from the perspective of *Synergetics* and its application to the domain of CTC (complexity theories of cities) by means of the notions of SIRN (synergetic inter-representation networks), IA (information adaptation) and their conjunction (SIRNIA). According to these theoretical perspectives, and as has been demonstrated in some details in Chap. 4 and subsequent chapters, urban dynamics is characterized by an on-going interaction between external information/data conveyed by the urban environment and internal information that originates in urban agents' mind/brain—a process captured by the notion of SIRN; in this process each urban agent adapts the incoming information/data from the city, by means of information previously constructed in the agent's mind/brain, as well as by the agent's chronic regulatory focus (i.e. its cognitive motivational inclination)—a process captured by the notion of IA.

Applied to the context of the present study, the external data/information is the size and properties of cities of various sizes as found by urban allometry studies, while the internal information is the promotion-prevention tendencies of the urban agents. The process of SIRNIA here refers to the way urban agents adapt to the information conveyed by the various urban environments, the effects of this information adaptation process on their motivation, action and decision making in the city and the resultant changes that take place in the city.

Our discussion below starts by suggesting a SIRNIA perspective on individual URF and on its role in the dynamics of cities (Sect. 13.2). In Part 3 we introduce a mathematical model constructed, on the one hand, on the basis of the theoretical foundation developed in previous chapters, while on the other, on the empirical findings of Ross and Portugali (2018). The chapter concludes with suggestions for further research.

13.2 A SIRNIA View on URF

The notion of SIRNIA, as developed in Chap. 4 and subsequent chapters, suggests that in the interaction between urban agents and their city, they are subject to two flows: A bottom-up flow of data that comes from the city and a top-down flow of information that originates in, and comes from, the agent's mind/brain/memory. We suggest that this process applies also to the experiments conducted by Ross and Portugali (cf. Sect. 13.1.3). Here, the bottom-up flow of data came not from the city itself, but from images of cities that were screened to the subjects. In line with our basic SIRNIA model, (Sect. 13.4), the brain first transformed the data into syntactic SHI, which triggered a top-down process that originated (1) in the person's previous knowledge, memory and experiences and (2) in the subject's chronic regulatory focus. Based on (1), the subject pattern recognized (and categorized) the screened city as fast-paced or slow-paced (thus transforming the SHI into SI) and then, based on an interaction between the pattern recognized image and (2), the subject's mind/brain adapted its chronic regulatory focus (that can be seen as a potential PI) to the pattern recognized fast-paced or slow-paced city, thus producing the materialized PI. In a real urban dynamics this will take the form of action and behavior in the city, whereas in the Ross and Portugali laboratory experiments (Sect. 13.1.3), this took the form of the subject's *response bias*. The latter measures the extent to which fast-paced and slow-paced cities affect the chronic promotion or prevention tendencies of the subjects.

According to Ross and Portugali's (ibid.) empirical findings, these PI adaptation processes, measured by means of the subjects' response bias, were as follows: when the city was pattern recognized as fast-paced, promotion oriented subjects tended to exhibit a more promotion-focused motivation/behavior (higher response bias), whereas prevention-oriented individuals tended to exhibit a more prevention-focused motivation and behavior (lower response bias). On the other hand, when the city was pattern recognized as slow-paced, those effects were eliminated. "Participants with a dominant promotion focus displayed responses that were equally conservative as displayed by those with a dominant prevention focus" (ibid., 9).

According to synergetics and by implication to SIRNIA, by means of their interaction with each other, the city's inhabitants bottom-up give rise to the emergence of a city's OP, that once comes into being, describes and prescribes ("enslaves") the behavior of the citizens in a top-down manner and so on in circular causality (cf. Chap. 11). Within this context, Ross and Portugali's (ibid.) empirical findings refer to the second, top-down aspect of the process—demonstrating that the city top-down affects the regulatory focus of its inhabitants and users. The question is how? We'll answer this question by developing a new urban model that explicitly theorizes about, and integrates RFT to the dynamic of cities.

13.3 Outline of the Model

From Ross and Portugali (ibid.) study follows that urban context of large, fast-paced cities encourages promotion-focused behavior; adding to this finding the logic of synergetics' circular causality process, it can be said that the encouraged promotion focus behavior intensifies the fast-paced dynamics of the city, which in turn reinforces and further strengthens a pattern of promotion-focused behavior, and so on and on in circular causality. This circular process may explain why in large, vibrant cities people take more risk, do more business, but also crimes, earn more money, produce and spend more.

As noted above (Sect. 13.1.3), in their laboratory experiments, Ross and Portugali have used the *response bias* as a measure of the extent to which fast-paced cities versus small-paced cities affected the personal (chronic) regulatory focus of the subjects. We denote the response bias variable by "b", where $b \geq 0$. Since we are dealing with an ensemble of citizens (represented by the test persons), we have to deal with the (relative frequency) distribution function

$$P(b;\ t) \tag{13.1}$$

where

$$\int_0^\infty P(b;t)db = 1$$

In their paper Ross and Portugali (2018) measure the mean $M \equiv \overline{b}$

$$\overline{b} = \int_0^\infty P(b;t)bdb \tag{13.2}$$

And the standard deviation σ (denoted by them by *SD*.) σ is defined as square root of the variance

$$S = \overline{b^2} - \overline{b}^2 \tag{13.3}$$

i.e. $\sigma = S^{\frac{1}{2}}$. Since our approach becomes much simpler, in terms of *S*, we will work with *S*.

How can we calculate *P* (13.1) or an equation for it in spite of the fact that we don't have any information on the processes on the "microscopic" level, e.g. social contacts between people etc., but only the "sparse data" (13.2) and (13.3)? To solve this problem, we resort to the method of making unbiased guesses on complex systems when only few data ("sparse data") are known (cf. Chap. 6). Starting from Jaynes' maximum (information) entropy principle and assuming a Markov process, a Fokker-Planck $\widehat{I}to$ equation of the distribution function *P*(*b*; *t*) can be derived (cf. Sect. 6.5). Here we use a form in which the drift and diffusion coefficients (or functions) are expressed as time-dependent functions (cf. (13.34) below).

Quite remarkably, we need not present equation (13.34) explicitly in the main text, because we may derive from (13.34) equations for the mean $\overline{b}$ (13.2) and the variance *S* (13.3), cf. (13.38) and (13.39), and perform our analysis on the level of these equations. After having provided the reader with this "background information" we start from the equations

$$\frac{d\overline{b}}{dt} = -\gamma\overline{b} + h(t) \tag{13.4}$$

$$\frac{dS}{dt} = -2\gamma S + 2Q(t) \tag{13.5}$$

where, for the moment being, the right hand sides are considered as hypotheses. To verify them and in particular to derive explicit expressions for *h*(*t*) and *Q*(*t*), we proceed in two steps (which, in a way, resemble a "gedanken" experiment). In the first step, we assume that we deal with people, whose behavior is not influenced by the city as a whole, but determined by their character and their reactions to typical random events. We assume, that their behavior can be modeled by a Fokker-Planck equation for which $h(t) = \gamma b_0$, and $Q(t) = Q$, where $\overline{b}$ is an average bias, whereas γ is the inverse time after which an individual returns to his/her "normal" state after some random incident. In the now following second step, we deal with the interplay between citizens and the city context so that we arrive at explicit expressions for $h(t)$ and Q. Here we must take care of the circular process described, in particular, at the beginning of this section. We consider the following two processes.

(1) The urban context U determines the behavior b, i.e. $P(b; t)$.

$$U \rightarrow P \tag{13.6}$$

In terms of SIRNIA, the "input" to this process is two information flows: the agent's 'chronic regulatory focus' and the information coming from the urban context; the "output" is the agent's response bias which emerges in the process of IA, that is, the interaction between the two input information flows.

(2) The behavior of the citizens determines the urban context U, i.e.

$$P \rightarrow U \tag{13.7}$$

In contrast to the process (13.6) referring to an individual, here we consider the *collective* output of all citizens giving rise to urban context and, interestingly, to a collective mental state ("attitude" in our case).

13.3.1 Impact of U on P

This impact can be taken care of by means of (13.4) and (13.5). U_M acts as additional "force" $F(t)$ on $\overline{b}$ in (13.4), so that we identify

$$F(t) = U_M \tag{13.8}$$

where, because of (13.7), U_M is a function or functional of $P(b; t)$. This dependence will be discussed below, Sect. 13.3.2. Since P is fixed by $\overline{b}$, and S, U_M can only depend on these quantities. According to Sect. 13.3.2, we may approximate U_M by

$$U_M \approx a_0\overline{b} + a\overline{b}^2 \tag{13.9}$$

There is also an impact of U on P via (13.5), or written as formula

$$\frac{dS}{dt} = -2\gamma S + 2Q + U_s \tag{13.10}$$

where, again, because of (13.7), U_S is function(al) of P, which we approximate by

$$U_s = c\overline{b} \tag{13.11}$$

where c is a constant. Thus (13.10) becomes

$$\frac{dS}{dt} = -2\gamma S + 2Q + c\overline{b} \tag{13.12}$$

13.3.2 Impact of P on U

As mentioned above, U_M and U_s must be functionals of $P(b;t)$ which can be expressed by $\overline{b}$ and S. In this way, U_M and Us become functionals of $\overline{b}$ and S at several times, i.e. of $b(t)$, $b(t')$ etc. (In fact, an individual adapts more quickly than a whole city).

We now make the assumption that the citizens adapt quickly so that we may neglect time delays such as $t - t'$ etc. In other words U_M, U_s become simple functions of $b(t)$, $S(t)$, e.g.

$$U_M = U_M(\overline{b}(t), S(t)) \tag{13.13}$$

Under the reasonable assumption that the impact of $\overline{b}(t)$, $S(t)$ is not too large, we may approximate U_M and U_s by low order polynomials of $\overline{b}$, S, where the impact of the size of $\overline{b}$ is far more dominant than that of S: An "observer" (new citizen) is far more impressed with the strength of activities (e.g. speed) rather than with their variety. Consequently

$$U_M \approx a_0\overline{b} + a\overline{b}^2 \tag{13.14}$$

When we insert (13.14) in (13.4) and (13.8) we observe that the term $a_0\,\overline{b}$ of (13.14) can be combined with the term $-\gamma\overline{b}$ in (13.4) so that, de facto, nothing has changed. Thus we may drop $a_0\,\overline{b}$ or just include it in $-\gamma\overline{b}$. All at all we arrive at our first fundamental model equation

$$\frac{d\overline{b}}{dt} = -\gamma\left(\overline{b} - b_0\right) + a\overline{b}^2 \tag{13.15}$$

We turn to U_s which is a measure of the restlessness, fluctuations, noise—in our case of a city/village. Here again, the size of $\overline{b}$ is the dominant cause, rather than the variation S of the distribution of b. Thus we arrive at our second fundamental model equation

$$\frac{dS}{dt} = -2\gamma S + 2Q + c\overline{b} \tag{13.16}$$

where it suffices to approximate U_s by a linear term c $\overline{b}$.

13.3.3 Solution to the Fundamental Equations

Equations (13.15) and (13.16) can be solved in two steps. First we solve (13.15) exactly. Then we insert the result in (13.16) which can be solved exactly (though perhaps (?) not in closed form). For our present purpose it suffices to treat the steady state solution to the Eqs. (13.15) and (13.16), i.e.

$$-\gamma(\overline{b} - b_0) + a\overline{b}^2 = 0 \tag{13.17}$$

And

$$-2\gamma S + 2Q + c\overline{b} = 0 \tag{13.18}$$

The quadratic Eq. (13.17) possesses two solutions (bifurcation!):

$$\overline{b}_1 = \frac{\gamma'}{2} + \frac{\gamma'}{2}\left(1 - \left(\frac{4}{\gamma'}\right)b_0\right)^{1/2} \tag{13.19}$$

$$\overline{b}_2 = \frac{\gamma'}{2} - \frac{\gamma'}{2}\left(1 - \left(\frac{4}{\gamma'}\right)b_0\right)^{1/2}, \tag{13.20}$$

where

$$\gamma' = \gamma / a$$

Concomitant with these solutions are S_1, S_2. By inserting $\overline{b}_1, \overline{b}_2$ into (13.18) we obtain

$$S_1 = \left(\frac{1}{2\gamma}\right)(2Q + c\overline{b}_1) \tag{13.21}$$

And, similarly

$$S_2 = \left(\frac{1}{2\gamma}\right)(2Q + c\overline{b}_2) \tag{13.22}$$

13.3.4 Making Contact with Observed Data

Our model contains 5 parameters, γ, b_0, a, Q, c. As it transpires from the form of (13.15) and (13.16), γ plays the role of a relaxation constant, or in other words, γ is the inverse of an adaptation time, τ. Since this is much shorter that the life span of an individual or even a city, we may assume that γ is large. Furthermore, we may measure the constants a, c, q In terms of γ. This entails that we may *formally* put

$$\gamma = 1 \tag{13.23}$$

(this means that we measure time in units $1/\gamma$).

So that we have to relate b_0, a, Q, c to observed data. We denote them by $\overline{b}_1(0), S_1(0), \overline{b}_2(0), S_2(0)$. Using (13.19), (13.20), (13.21), (13.22) and (13.23) we readily obtain

$$\frac{1}{a} = \overline{b}_2(0) + \overline{b}_1(0) > 0 \tag{13.24}$$

$$a\overline{b}_1(0)\overline{b}_2(0) = b_0 \tag{13.25}$$

And

$$c = 2(\overline{b}_1(0) - \overline{b}_2(0))^{-1}(S_1(0) - S_2(0)) \tag{13.26}$$

$$Q = (S_1(0) + S_2(0)) - (S_1(0) - S_2(0))(\overline{b}_2(0) + \overline{b}_1(0)(\overline{b}_1(0) - \overline{b}_2(0))^2 \tag{13.27}$$

It remains to discuss how to obtain the data of the mean and standard deviation σ (or equivalently $S = \sigma^2$) of the response bias.

(a) By means of field experiments with populations of large/small cities.
(b) As studies in Ross and Portugali paper (cf. Sect. 13.1.3) by "laboratory" experiments.

Here we use the latter approach and analyze its data in the light of our model.

We observe that the relations (13.24), (13.25), (13.26) and (13.27) allow us to calculate the model parameters a, b, c, Q from measured data, while, conversely, the relations (13.19), (13.20), (13.21) and (13.22) allow us to "predict" the observed data on the basis of the model parameters. To illustrate our further procedure we analyze "the effect of urban context on *promotion-focused* participants" (Ross and Portugali 2018). We denote this case by (1a) promotion low.

According to Ross and Portugali (ibid.).

Small city: mean M = 0.15, standard deviation $\sigma = 0.11$

Big city: M = 0.11, $\sigma = 0.11$

Table 13.1 Calculated parameters b_0, a, c, Q

Case	b_0	a	c	Q	$\overline{b}_1$	$\overline{b}_2$	S_1	S_2
1a	0.064	4	0	0.01	0.15	0.11	0.001	0.001
1b	0.084	3	0.06	−0.004	0.15	0.18	0.014	0.017
2a	0.087	3	0.8	−0.052	0.16	0.18	0.012	0.020
2b	0.064	4	0.2	−0.001	0.15	0.11	0.014	0.010
3a	0.09	2.8	0.3	−0.01	0.16	0.2	0.014	0.02
3b	0.051	4.6	1.2	−0.042	0.14	0.08	0.12	0.008

We denote the mean of small city by $\overline{b}_1$ and of big city by $\overline{b}_2$. We use the variance $S = \sigma^2$ with indices 1, 2 corresponding to $\overline{b}_1, \overline{b}_2$. Thus

$$\overline{b}_1 = 0.15, \quad \overline{b}_2 = 0.11, \quad S_1 = 0.012, \quad S_2 = 0.012$$

Using (13.24)— (13.27) we obtain

$$b_0 = 0.064, \quad a = 4, c = 0, \quad Q = 0.01$$

We apply this procedure to all cases reported by Ross and Portugali (2018):

promotion focus low, (1b) high
prevention focus low, (2b) high
dominant focus: promotion, (3b) prevention.

Our results are listed in Table 13.1.

While the results on M, σ (or in our notation $\overline{b}, S$) were discussed by Ross and Portugali, scrutinizing Table 13.1 leads us to some new insights based on the parameters b_0, a, c. First of all, $c > 0$ means that larger bias b leads to higher fluctuations in accordance with our above hypothesis. The result that in all cases $\overline{b}_1 > b_0, \overline{b}_2 > b_0$ means that living in a community enhances the personal bias; however, differently depending on his/her regulatory focus. Concerning b_0 and a we notice a remarkable symmetry between (1a) promotion focus low, and (2b) prevention focus high. This means, that low promotion focus and high prevention focus have the same effect on $\overline{b}_1, \overline{b}_2$. On the other hand, the fluctuations of S are smaller in case (1a) than in (2b).

Approximately the same symmetry holds in the cases (1b), (2a), but this time including the fluctuations. In the case (3a) b_0 is still somewhat larger than that of case (1b) as can be expected. We may interpret ab_0 as "effective" social interaction "constant". And, in fact, in all cases (1a) till (3b) this product yields practically the same value!. While our model captures qualitatively and quantitively how the city context changes the mean bias of persons and the variance/standard deviation, we are aware that it was based on experimental data of 201 graduate students from the University of Tel-Aviv. Surely, more data and studies are thus needed to validate the model and to draw more general conclusions about behavior of persons living in large and small cities.

13.3.5 Stability of Solutions

To bring out the essentials, we consider (13.15) for the special case $b_0 = 0$, i.e.

$$\frac{d\overline{b}}{dt} = -\gamma\overline{b} + a\overline{b}^2 \tag{13.28}$$

The steady state solutions are

$$\overline{b}_2 = \gamma/a \tag{13.29}$$

$$\overline{b}_1 = 0 \tag{13.30}$$

To check their linear stability, we insert $\overline{b}_2 = \frac{\gamma}{a} + \varepsilon, \overline{b}_1 = \eta$ in (13.30) and obtain (in linear approximation)

$$\frac{d\varepsilon}{dt} = \gamma\varepsilon \tag{13.31}$$

And

$$\frac{d\eta}{dt} = -\gamma\eta \tag{13.32}$$

While $\overline{b}_1$ represents a stable solutions, $\overline{b}_2$ is unstable. What does this imply? To this end, we rewrite (13.28) by means of a potential V in the form

$$\frac{d\overline{b}}{dt} = -\frac{\partial V}{\partial \overline{b}}, V = \gamma\frac{\overline{b}^2}{2} - \frac{a}{3}\overline{b}^3 \tag{13.33}$$

Figure 13.3 shows a plot of V. In terms of Fig. 13.3, the position of the ball on the slope of a hill symbolizes the size of $\overline{b}$ and its further development. Evidently the ball on the right slope of Fig. 13.3 *left* indicates that $\overline{b}$ will increase indefinitely. i.e. that the average response bias increases forever—surely in contrast to human nature. Thus in order to apply our model to a real situation we must invoke a barrier such as drawn in Fig. 13.3, *right*. Mathematically, the barrier can be taken care of by an additional term in (13.33), whose explicit form is not important for our discussion. As can be seen, $\overline{b}_2$ is a lower limit for people who want to belong to the "large city" group.

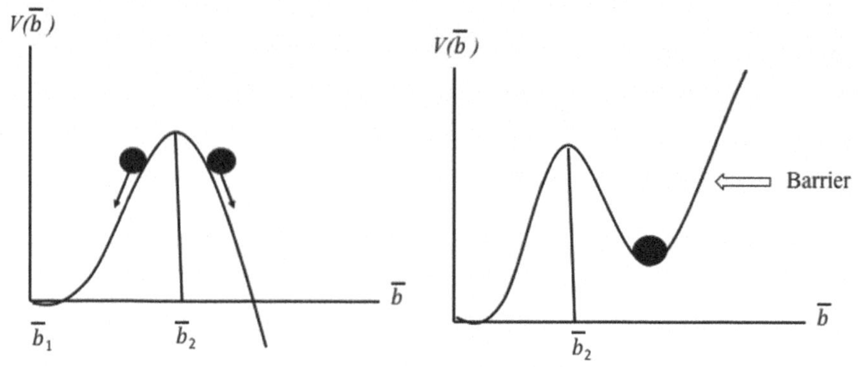

Fig. 13.3 Left: potential "landscape". Right: Potential "landscape" with a barrier

Note, firstly, that in the light of these considerations and the results of Ross and Portugali's (Sect. 13.1.3), a high enough promotion focus seems to be a necessary prerequisite for the formation of the "large city" group. Secondly, that the just mentioned behavior of $\overline{b}$, that can be read from Figs. 13.3 *left* and *right*, is fully substantiated by the exact time-dependent solutions to Eq. (13.15) (including $b_0 \neq 0$).

13.3.6 Summary of the Model

In our above model, we have dealt with the following phenomenon that results, as noted above, from Ross and Portugali study: Urban context of large, fast-paced cities encourages promotion-focused behavior; adding to this empirical finding the logic of synergetics' circular causality process, it can be said that the encouraged promotion focus behavior intensifies the fast-paced dynamics of the city, which in turn reinforces and further strengthens a pattern of promotion-focused behavior, and so on. Thus we are confronted with a problem of circular causality in a system with many participants, a problem at the core of Haken's (1983) synergetics (cf Chap. 3). In the parlance of synergetics, $\overline{b}$ plays the role of an order parameter and (13.15) is the order parameter equation. To tackle this specific problem, following a suggestion made in Ross and Portugali (2018), we have chosen the "response bias" of people, b, as the characteristic variable. Our central task has it been, to determine the equation for the probability distribution function $P(b; t)$ that develops in the course of time, t, because of the circular process, and solve this equation. To this end we have performed the following steps:

1. Since we aim at a comparison with measured data, that are the mean value M and standard deviation σ and that are as compared to the complete distribution function $P(b; t)$ *sparse data*, we invoked a method that allows us to make the best guess on an equation for $P(b; t)$ under the conditions of given $\overline{b}$ and σ

or S (Haken 2004) (for a short summary cf. the appendix). This leads us to a Fokker–Planck—$\hat{I}$to equation.
2. Instead of solving this partial differential equation, we derive from it equations for b and S, first for people not in urban context.
3. We extend the $\overline{b}$ and S equation by taking the urban context into account.

At each step we make well-defined and well-justified approximations so that our final model equations rest on safe ground. In accordance with observed data, we obtain two distinct solutions corresponding to the large and small city cases. In particular we have found in accordance with Ross and Portugali (2018), that a high chronic promotion focus is a precondition for the formation of a "large city" group.

13.4 Conclusions

Our aim in this chapter was to study the way humans' basic motivational-behavioral tendencies are related to the dynamics of cities as complex, adaptive, self-organization systems—the ways the dynamic of cities of different sizes is affected by, and is affecting, the promotion and prevention tendencies of their inhabitants and users. We explored these issues from the theoretical perspectives of Synergetics approach to complexity theories of cities through the notions of SIRN, IA and their conjunction (SIRNIA). From these theoretical perspective we firstly suggested a descriptive account and then a mathematical model; both illustrate how the circularly causal process by which urban contexts of large, fast-paced and slow-paces cities affect, and are affected by, the promotion- and prevention-focused behavior of their citizens. As specified above, the urban process associated with this circularly causal process applies to both very large and very small cities and towns, and includes the various ingredients of cities as complex systems: self-organization, a bottom-up emergence, and, qualitative phase-transition changes. Note that our approach sheds light on the creativity of evolutionary processes, in which people's error-making, mutations and innovations, are also a result of evolution.[1] While interesting and significant, these issues should await a subsequent study.

Finally we should reiterate here the generalization we made in Chap. 11 Sect. 11.2, namely, that employing URF is but one way to deal with human behavior; there are other cases of behavioral patterns/features such as the walking speed of pedestrians that we discussed in Chap. 6, Sect. 6.5. Here too, further elaboration on this issue will have to await a subsequent study.

[1] We thank an anonymous reviewer for this comment.

Appendix

Basic equations

Wanted: An equation for the best fit of a probability distribution function *P*(*b*;*t*), of which only mean and standard deviation are known.

Answer: Making an "unbiased guess under constraints" allows us to derive a Fokker–Planck—$\hat{I}$to equation for *P*(*b*;*t*), which in the present special case reduces to the Fokker-Planck equation

$$\frac{dP(b;t)}{dt} = -\frac{d}{db}((-\gamma b + h(t))P(b;t)) + Q(t)\frac{d^2(P(b;t)}{db^2}) \tag{13.34}$$

where *h* and *Q* are independent of *b*.

In the absence of urban context, we may put $h(t) = \gamma b_0$, $Q(t) = const.$

As we derived in the main text, in the case of urban context, we may put

$$h(t) = \gamma b_0 + U_M,\ Q(t) = Q + \frac{1}{2}U_S \tag{13.35}$$

where most importantly, U_M, U_S are constants independent of b (but depending on $\overline{b}$). Thus, U_M, U_S don't change the structure of (13.34). This equation allows us to derive equations for the mean

$$\overline{b} = \int_0^\infty bP(b;t)db, \tag{13.36}$$

And variance

$$S = \int_0^\infty b^2 P(b;t)db - \overline{b}^2 \tag{13.37}$$

from which we may calculate the standard deviation σ by $\sigma = S^{\frac{1}{2}}$. To this end we multiply (13.34), by *b* or b^2, respectively, and integrate both sides over *b*. After partial integrations we readily obtain.

$$\frac{d\overline{b}}{dt} = -\gamma\overline{b} + h(t), \tag{13.38}$$

$$\frac{dS}{dt} = 2\gamma S + 2Q(t) \tag{13.39}$$

References

Bettencourt, L. M. A. (2013). The origins of scaling in cities. *Science, 340,* 1438–1441.

Bettencourt, L. M. A., Lobo, J., Helbing, D., Kühnert, C., & West, G. B. (2007). Growth, innovation, scaling, and the pace of life in cities. *Proceedings of National Academy Science U.S.A., 104*(17), 7301–7306. https://doi.org/10.1073/pnas.0610172104.

Chen, Y. (2021). Scaling, Fractals, and Spatial Complexity of Cities. In J. Portugali (Ed.), *Handbook on Cities and Complexity E Elgar Publishers.* England.

Crowe, E., & Higgins, E. T. (1997). Regulatory focus and strategic inclinations: Promotion and prevention in decision-making. *Organizational Behavior Human Decision Processes, 69,* 117–132. https://doi.org/10.1006/obhd.1996.2675.

Faddegon, K., Scheepers, D., & Ellemers, N. (2008). If we have the will, there will be a way: Regulatory focus as a group identity. *European Journal Society Psychology, 38,* 880–895. https://doi.org/10.1002/ejsp.483.

Haken, H. (1983). *Advanced synergetics.* Berlin/Heidelberg/New York: Springe.

Haken, H. (2004). *Synergetics. Introduction and advanced topics.* Springer: Berlin, Heidelberg.

Higgins, E. T. (1997). Beyond pleasure and pain. *American Psychologist, 52*(12), 1280–1300.

Higgins, E. T., Friedman, R. S., Harlow, R. E., Idson, L. C., Ayduk, O. N., & Taylor, A. (2001). Achievement orientations from subjective histories of success: Promotion pride versus prevention pride. *European Journal of Society Psychology, 31,* 3–23. https://doi.org/10.1002/ejsp.27.

Milgram, S. (1970). The experience of living in cities. *Science, 167,* 1461–1468.

Ribeiro, F. L., Meirelles, J., Ferreira, F. F., & Neto, C. R. (2017). A model of urban scaling laws based on distance dependent interactions. *Royal Society Open Science, 4*(3), 160926.

Ross, G. M. (2015). *Urban regulatory focus: Promotion and prevention are part of the identity of the city.* PhD dissertation, TAU.

Ross, G. M., & Portugali, J. (2018). *Urban regulatory focus: A new concept linking city size to human behavior.*

Simmel, G. (1964). The metropolis and mental life. In W. K. Simmel (Ed.), *The sociology of George* (Part Five, Chap. IV, pp. 409–424). New York: Free Press.

Wirth, L. (1938). Urbanism as a way of life. *The American Journal of Sociology, 44*(1), 1–24.

Chapter 14
Smart Cities: Distributed Intelligence or Central Planning?

14.1 Introduction

> Cities were always smart. In every era, advanced technologies and innovative thinking have developed in cities; from the written word 5000 years ago; to the revolutionary Greek concepts of democracy and citizenry; to Renaissance art and architecture; to the factories of the industrial revolution; to today's post-industrial age of high technology. (Portugali 2016a)

This leads us to the question whether there is a specific feature that distinguishes the present days' concept of a "smart city" (Batty et al. 2012) from those of history. We think that a clue to answer this question is a look at our present most frequent use of the word "smart ". In fact, we speak of smart phones, but also of smart houses, smart households, smart cars etc. At a still larger scale, smart factories are conceived. These smart objects are related to the notion of *The Fourth Industrial Revolution* (Schwab 2016) suggesting that today once again society is at the threshold of an industrial revolution: The 1st happened in the eighteenth century by the introduction of mechanical machines, the 2nd in the 2nd half of the nineteenth century via electrification. The 3rd revolution is characterized by computers and microelectronics, while presently we witness the rise of smart factories by what is called "digitization" of a network comprising construction, development, production, sales, and services (cf. Bauernhansl 2015). This implies an integration between real and virtual worlds, enabling the simulation of systems, processes and even complete factory plants in real time. Here, the reconciliation between long term planning and short term reactions to customers' wishes and market fluctuations presents a real challenge. Planners presently consider the human–robot cooperation from the point of view of a combination of the cognitive superiority and flexibility of humans and the power, endurance and reliability of robots. Quite evidently, the concept and development of smart cities will present a far greater challenge that has to take care of the central task of a city—the welfare of its citizens.

H. Haken and J. Portugali, *Synergetic Cities: Information, Steady State and Phase Transition*, Springer Series in Synergetics,
https://doi.org/10.1007/978-3-030-63457-5_14

The currently developing discourse on smart cities is intimatly associated with CTC (complexity theories of cities) that we've discussed in Chap. 2. Smart cities with their massive use of AI/IT are considered as one central way to cope with the growing complexity of cities, while the set of urban simulation models developed within the context of CTC is considered one among the new ICT (information communication technology) that enables the smartification of cities (Batty et al. 2012).

As we indicate in Chap. 2, while cities share many properties with natural complex systems, they differ from the latter in that they are *hybrid complex systems* composed of artifact that are by definition simple systems and of human urban agents that are natural complex systems. In today's cities there is a clear distinction between the artificial components of cities as hybrid complex systems (e.g. houses, streets, etc.) and their natural components—the human urban agents. Artifacts cannot interact (exchange things/information), agents can. Artifacts interaction is thus *mediated* by agents. Also, a lot of agents' interaction is mediated by artifacts (cf. Chap. 4). With AI/IT, artifacts might be able to interact directly with each other without mediation by human agents and thus become *artificial urban agents.* E.g. a self-driven car can interact with a self-organized traffic light in a junction. Artifacts thus become urban agents giving rise to a new form of urban dynamics. Part of it may be smart traffic regulations (guiding system), smart supply systems of energy, food, goods, or smart waste disposal. But nevertheless: Who or what is smart?

The technical installations (houses, communication systems, supply systems), their human planners and users, or their interplay? At any rate, we are facing a delegation of human decision making and responsibilities to automata. So all in all we believe that "smart city" implies a qualitative change based on AI (artificial intelligence) embodied by IT (information technology) on all scales. This means that we must not ignore the fact that cities are embedded in nations in an increasingly connected global world which is becoming also "smart" using AI (just think of the financial market with its computer controlled transactions). At all levels, there will be a demand for sensors, actuators and computing power. The wide use of AI/IT devices will lead to an innovation wave quite welcome to economists because innovations are seen as a motor of economic growth and public welfare. In the following we want to elucidate some aspects of "smartification".

Basically, we may distinguish between two approaches: The top down approach, where data are collected locally and sent to a central computer, which makes the decisions, or a bottom up approach where the decisions are made at the individual level based on collected data. The latter approach is outlined in a study by Feder-Levy et al. (2016), based on Portugali's concept of selforganizing city (Portugali 2000, 2011, 2012). As we've seen in previous chapters (specifically Chaps. 3, 5 and 6) the theory of Synergetics offers an integrative approach: Local bottom–up decisions and actions give rise to a collective structure that then in a top–down manner determines ("enslaves") local actions and decisions and so on in circular causality.

In the beginning, solid structures such as houses, streets, will probably not be changed. Only at a later stage, changes may become necessary due to a newly developing dynamics of traffic, but also of personal habits. E.g. as is evident in the new reality of the corona pandemic, IT has the potential to lead to a delocalization of teaching and, perhaps, education, by replacing schools and universtities by tele-courses, e.g. MOOC (massive open only courses). Because this implies a loss of personal contact between teachers and students (and among students) it is, however, unlikely that schools and universities will disappear completely. The same holds true for sports—or cultural centers. Can e.g. in a baseball or soccer game virtual reality replace the feeling of being member of some community?

14.1.1 On the Interplay Between Humans and Smart Devices

Smart devices on nearly all scales will replace human senses and (re)actions. A few examples may illustrate this. In a smart home sensors may measure the actual amount of daylight/sunshine, rain fall, in- and outdoor humidity and temperature, energy consumption etc. In a city, sensors may measure local traffic flows, local and overall energy consumption etc. So far in a home, individuals have reacted to the data, e.g. temperature according to their specific habits. Now, "AI-devices" learn all these human reactions, e.g. to close the venetian blinds at a certain level of sunshine. But there may be conflicts between family members. While A wants them to be closed, B wants them to remain open. This learning problem can perhaps be solved by some kind of majority decision based on relative frequences of action. At any rate the AI device will then act instead of a person. This is surely convenient for the individual, but leads also to a reinforcement of his/her habits, i.e. his/her habituation.

From the point of view of the complexity theory of Synergetics, the above situation implies that the AI program has become the order parameter (OP) that now, in the literal sense of the word, enslaves the individuals. Having the properties of OPs in mind, it will be difficult to change that OP. Such habituation effects have been repeatedly discussed in connection with advertisments based on consumers' behavior. At the political level, even some kind of nudging has been discussed.

A down to earth technical problem should be mentioned when WLAN is used in neighboring flats/homes, interference effects may spoil the operations. Both, *conflicts* of interests among citizens as well as habituation may occur at the city level: chosen car routes (think of autonomous cars interacting with traffic guidance systems), total energy consumption, use of communication channels, supply routes for commodities etc.

While in this way, the former more or less self-organized collective behavior of the citizens will be learned by the AI system and "regulated" correspondingly we may think of quite another scenario: a central city computer tries to solve a multi-travelling salesman problem. Though we don't know how such an approach would look like or what its results/efficiency will be, such a scenario is by no means unlikely. This may perhaps lead to large computer centers outside cities where building sites are cheap. Finally, centralized installations will increase the vulnerability of cities against crime, terror acts and global break down. Other basic problems are, e.g. energy consumption control. A typical "recipe" runs like this: increase/decrease of energy price over the day/night depending on consumption thus influencing the consumers' action. As is well known this scenario may lead to instabilities in the network. A stricter central control may eliminate the instabilities, but will curb individual freedom: clearly, smartification has sociological implications.

14.1.2 Theoretical Tools

To study implications of "smartification" from the point of view of theory, we have a number of approaches at hand, i.e.

(1) Cognitive Science
(2) Artificial intelligence
(3) Information theory
(4) Synergetics
(5) Network theory
(6) Multi-agents theory
(7) Evolutionary game theory
(8) Allometry/Scaling laws
(9) Biology—Evolution, population dynamics

Besides these approaches with their roots in mathematics and the (natural) sciences, other disciplines will play a role in smart cities such as.

(a) jurisprudence (e.g. responsibility transfer, liability)
(b) sociology (e.g. job market requiring highly qualified personal for IT-maintenance, development of AI/IT, but also disappearence of other jobs, e.g. taxi drivers, bank clerks etc.)
(c) psychology (e.g. psychological stress in an automated world)
d) political science (e.g. decision making on cities' smartification in a democratic society)
(e) economics (e.g. investments in smartification)
(f) ecology (e.g. management of resources by smartification)

Having said this, we focus our attention on some of the topics 1–9. First we briefly discuss them. Since, at least according to our understanding, "smart city" implies a massive use of AI/IT, a discussion on intelligence in general and on AI in particular may be in order. Here, we have to draw on insights gained by Cognitive Science. We will elaborate this in Sect. 14.2. In this context, we will briefly discuss information theory and its more recently established connections with cognition such as *information adaptation* (cf. Chap. 4).

Cities with their inhabitants, and mobile and immobile installations (artifacts) are truely complex systems that form specific spatial and functional structures where the interplay or competition between central planning and local personal initiative becomes quite decisive (Chaps. 2 and 15). A general theory of structure formation in complex systems is provided by Synergetics that we have outlined in previous sections. This will shed light on the role of indirect steering in contrast to conventional planning. Network theory (cf. e.g. Barabási 2016) gives insight into the links, e.g. between inhabitants or between neighborhoods or cities, and allows, e.g., the derivation of scaling laws. On the other hand, multi-agent theories (cf. e.g. Bretagnolle et al. 2006; Roscia et al. 2013) deal with the actions of citizens and automata. Quite clearly, network theory will play an important role in IT dealing with communication and transport. In the context of this chapter we are rather concerned with AI aspects. Evolutionary game theory (EGT) deals with the benefits of cooperation between partners (persons, companies, institutions etc.). The concept of Nash equilibria (well known in EGT) will play a role in our discussion. EGT originated from game theory developed by von Neumann and Morgenstern (1944). The biological concept of evolution, in particular based on *mutations*, has become an important ingredient of EGT (e.g. Nowak 2006). Mutations are treated as chance events, also denoted as *fluctuations*. As is shown quite generally in Synergetics, fluctuations play a fundamental role in the selforganized formation of structures. Fluctuations can be also considered as means to test the stability or resilience of a system against perturbations and to trigger novel developments. A further line of research inspired by biology is *allometry* which is being applied to theories on city growth. In biology it was found that e.g. the age, size, blood flow and other characteristic features of animals scale with body weight w by means of a power law, $\propto w^{l/4}$, $l = 1, 2, 3$ (West and Brown 2005). Similar laws have been considered in the case of cities (Bettencourt et al. 2007). We will discuss such laws with respect to "smart city" at several instances below (Sect. 14.5).

14.2 Intelligence

Having in mind that "smartification" implies the application of AI to cities, a few remarks on intelligence may be in order, at least what its role in cities concerns. Most probably, it is sufficient to deal with *intelligent behavior*. This in turn means: appropriate *actions*, or depending on the specific situation, *reactions*. A prerequisite for these processes is *pattern recognition*, where "pattern" may be interpreted in a wide sense, e.g. as images of faces, or objects, or whole scenes, or movement patterns of persons, groups of them, of traffic, or just a set of measured data, e.g. temperature distribution in a home or in a city. The central problem consists in the selection/recognition of features that are/will be relevant for (re)action and to draw the relevant conclusions. To have the required actions *quickly* at hand, a whole repertoire of (re-)actions must be learned in advance. A number of them may be based on planning, and foresight is required. An important problem is the challenge of new situations to decision making. In many cases, people base the latter on more or less justified analogies with previous experience by extrapolating ("heuristics"). But in a number of cases, entirely new solutions (actions) are required. Among such chance events are technical failures, human mistakes, natural catastrophes, but even new state laws etc.

So far we have spoken of intelligent behavior of an individual. But there is also collective intelligence. The famous economist Friedrich August von Hayek spoke of "intelligence of the market" which in the present context can be extended into the "intelligence of the city". Some collective intelligence is even attributed to swarms of birds or schools of fish. Quite a number of scientists think that collective intelligence cannot be substituted by individual intelligence. At least one of the reasons is the "information bottleneck": a central agency, or even an individual, cannot deal with the huge amount of incoming information. Thus only delocalized decision making (by the market) is possible. Clearly, there is a dichotomy between local initiative and global planning. Coming back to "smartification". Taken seriously, it means the replacement of human intelligence—as sketched above—by artificial, that is, machine, intelligence.

14.2.1 Artificial Intelligence (Machine Intelligence)

Here we don't discuss philosophical issues such as the "ghost in the machine" nor ethical, such as responsibility or liability of machines. But cf. e.g. Bonnefon et al. (2016).

The first step consists in the collection of data by sensors for optical, acoustic, chemical, tactile, acceleration etc. signals. Present days' data may be enormous; hence the notion of "big data". If not appropriately preselected ("filtered") they mirror a complex world. The next, in our view decisive step is supervised learning (cf. e.g. Le Cun et al. 2015). For example, in image recognition, a huge number of faces of the same person but in different positions, under different illuminations is presented to the computer that has to "learn" that all these images are *associated* with the name of that person. This is accomplished by an algorithm with a huge number of adjustable parameters. This procedure is usually visualized as a feed-forward net with a considerable number of layers (say 20–100). Most of the adjustment procedures are based on the method of "steepest descent". To get an idea what this means and implies, think of a landscape with mountains and valleys in between. The bottom of each valley represents a specific prototype pattern to be recognized due to given data which are represented by the position of a stone in this landscape. Since in general the data are incomplete or/and partially erroneous, the position of the stone does not precisely coincide with that of the bottom of the relevant valley. But the stone can correct this error by sliding downhill.

This little side remark may shed light on a basic problem: how certain is it, that the "envisaged" valley is the correct one? This also implies that under somewhat changed conditions, the previous valley is no more the appropriate one if it has happened so before. This means there is no guarantee for the adaptability or generalization capability of this algrorithm. Besides such a rather superficial consideration (and related ones) there is no theory of such networks that would allow us to understand the learning process. In a way the whole learning procedure is reminiscent of the training of animals. We don't know, what happens in their brains. And perhaps, once the tiger bites, quite unexpectedly. An additional remark may be in order: at least at present, the training time of such a network is days, which may be shortened by the parallel use of thousands of computers. Nevertheless, the speech recognition capability of systems such as Siri (Apple) is impressive. So far we have talked about recognition (which in case of AI may include that of scenes and of actions of persons). What about action of automata? The same way as recognition can be "taught" to a computer, it can also be taught to steer movements of actuators in particular by *imitation*. But then, quite often, the machine is confronted with a conflict situation: which movement to steer (perform)? The unability of the machine to make an autonomous decision leads to deadlock. In human life, in such a situation a deeper insight, or deeper experience is required and helps to overcome the deadlock. On the other hand, if such a conflict situation hasn't been "shown" to the machine—deadlock results.

14.3 Information Dynamics and Allometry in Smart Cities

In Sect. 10.5 we have dealt with information production (and processing) in "traditional" cities, based on the notions of SHI, SI and PI. These notions and their relations in connection with cities, were introduced in Chap. 4 in the context of SIRNIA. Here our aim is to apply them to the smartification of cities. We briefly remind the reader of our approach of Sect. 10.5. Humans ("agents") play a double role: they receive information and produce information. We assume that the received information is of Shannon type and the "produced" information of pragmatic type. The latter leads to observable actions of the agents. As we show in some detail in Chap. 4, because of their recognition capabilities the agents transform the data emitted by the environment firstly to SHI and then into PI. For simplicity, in what follows we'll deal with the conversion of SHI into PI. Let us consider the role of the city that is composed of artifacts and agents. Both emit signals (e.g. the artifacts in Gibsonian language as affordances). All these signals are treated as "raw material" in form of SHI that has then to be "deciphered" by the agents i.e. converted in PI.

Thus SHI is the number of bits produced by a city, e.g. per day, and embodied by letters, on monitors—a gigantic, but nevertheless measurable number. These signals are received, filtered and interpreted by the human sensorium and eventually converted into PI. This leads us to our model depicted in Fig. 14.1a. As can be seen, Fig. 14.1 illustrates graphically four stages in the smartification of a city, starting from Fig. 14.1a, representing the current state of cities, all the way to Fig. 14.1d that represents a "fully smart" city as discussed in more detail, below. The four stages in the smartification of a city are:

(a) The city and its human inhabitants exchange information. While the city produces SHI, its human inhabitants recognize these signals and convert it via PI into actions
(b) Automata participate in the exchange process

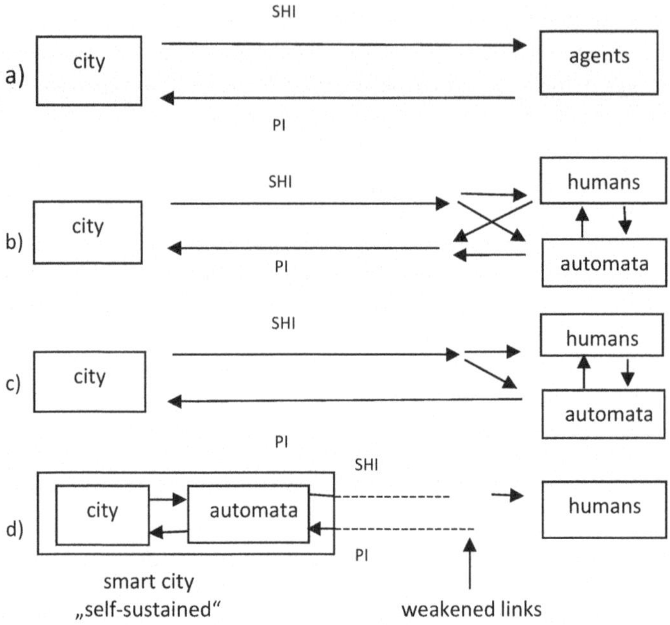

Fig. 14.1 *Stages of smartification (cf. also text).In the case of automata dominance* $a^2 r_a r'_a/\gamma_a >> n^2 r_h r'_h/\gamma_h$ (*cf.* Sect. 14.6)

(c) The direct action of humans on the city has practically disappeared and is mediated by automata
(d) Automata are perceived by humans as part of the city.

Now we focus our attention on the role played by automata. Our basic approach is depicted in Fig. 14.1. We distinguish between three producers and processors of information: the city as a whole comprising artifacts (buildings, streets, etc.), human agents, and automata (artificial agents). We treat the city as a sender of information (where the role of artifacts is interpreted in Gibson's sense of affordances). According to SIRNIA (Chap. 4, Fig. 4.9), the emitted information is treated as SHI and measured in bits. Its recognition is left to the receivers, i.e. to humans and automata. Thus according to our previous analysis, SHI is converted into PI (and SI). PI becomes visible as specific actions performed hy humans and automata. Actually, pattern recognition capabilities are ascribed to the latter, as well as the capabilities of appropriate reactions. We include SI (semantics) in PI, provided ideas, concepts, plans, etc. ("mental states") are externalized, e.g. by written texts, computer programs, including those for graphics, etc. (see also the concept of SIRN). We measure PI in bits. Since in our approach we are dealing with grossfeatures to illuminate the impact of automata on information we consider the total amount of SHI and PI per 24 h in a city. We assume an unlimited channel capacitiy. Clearly, more IT oriented approaches will have to consider the effect of channel capacity also. We assume that the conversion of SHI→PI occurs within 24 h ("temporal course graining") and we ignore the local fine structure ("spatial course graining"). We treat the city as an open system with imports and exports of raw material, commodities, ideas etc. The variables of our basic equations are

SHI produced per day by the city, denoted by s.
PI produced per day by humans, denoted by p_h.
PI produced per day by automata, denoted by p_a.
n is the number of citizens.
a is the number of automata.

In the spirit of Synergetics, s, p_h, p_a are the order parameters while n and a act as control parameters jointly with rate constants defined in the following.

We come to the formulation of rate equations

p-rates

(1) generation rate of p_h (by human actions) $\frac{dp_h}{dt}|_1 = n \cdot s \cdot r_h$: humans transfer incoming Shannon information into PI_h (actions) at rate r_h
(2) generation rate of p_a by automata $\frac{dp_a}{dt}|_1 = a\, s\, r_a$: automata transfer incoming Shannon information into PI_a (actions) at rate r_a
(3) loss rate of p_h $\frac{dp_h}{dt}|_2 = -\gamma_h p_h$
(4) loss rate of p_a $\frac{dp_a}{dt}|_2 = -\gamma_a p_a$ loss because of forgetting, executed actions by humans or automata, storage in memory.

s-rates SHI is produced by the city

(5) generation rate of s $\frac{ds}{dt}$ = generated data D
(6) by humans $D_h = n \cdot g_h + n\, r'_h p_h$, g_h: spontaneous generation rate per human, $r'_h p_h$: stimulated actions per human
(7) by measurement devices D_a(automata) e.g. energy consumption, traffic flow $D_a = a g_a + a r'_a p_a$, g_a: generation rate per automaton, $r'_a p_a$: stimulated action per automaton
(8) loss rate of s:
(9) by conversion $s \to p_h$:$\tilde{r}_h s n$ (via humans)
(10) by conversion $s \to p_a$:$\tilde{r}_a s\, a$ (via automata)
(11) by errors, $\gamma_e\, s$, where γ_e error rate
(12) by spam, chunk included in (11) total rate s
(13) $\frac{ds}{dt} = n \cdot g_h + a\, g_a + n\, r'_h p_h + a r'_a p_a - \tilde{r}_h s\, n - \tilde{r}_a s\, a - \gamma_e s$
We consider the steady state
(14) $\frac{dp_h}{dt} = \frac{dp_a}{dt} = \frac{ds}{dt} = 0$
The sum of the generation rate (1) and loss rate (3) jointly with (14) yields
(15) $n\, s\, r_h - \gamma_h p_h = 0$
Similarly, (2), (4), (14) yields
(16) $a\, s\, r_a - \gamma_a p_a = 0$
The sum over (6), (7), (9), (10), (11), jointly with (14) yields

(17) $n\, g_h + a\, g_a + n\, r'_h p_h + a r'_a p_a - \tilde{r}_h s\, n - \tilde{r}_a s\, a - \gamma_e s = 0$

These three equations for s and p_h, p_a depend, in particular, on the parameters n (humans) and a (automata).

Solution to (17):

(18) $s = (\tilde{r}_h n + \tilde{r}_a a + \gamma_e)^{-1}\left(n\, g_h + a\, g_a + n\, r'_h p_h + a\, r'_a p_a\right)$

Solution to (15) and (16):

(19) $p_h = \gamma_h^{-1} n\, s\, r_h$

(20) $p_a = \gamma_a^{-1} a\, s\, r_a$

from which we obtain

(21) $p_h / p_a = \gamma_h^{-1} n\, r_h / \left(\gamma_a^{-1} a\, r_a\right)$

or, equivalently

(22) $p_a = A p_h,\ A = \gamma_a^{-1} a r_a / \left(\gamma_h^{-1} n r_h\right)$

and with (18)

(23) $p_h = \gamma_h^{-1} n\, r_h (\tilde{r}_h n + \tilde{r}_a a + \gamma_e)^{-1}\left(n\, g_h + a\, g_a + n r'_h p_h + a r'_a p_a\right)$

(24) $p_a = \gamma_a^{-1} a\, r_a (\tilde{r}_h n + \tilde{r}_a a + \gamma_e)^{-1}\left(n\, g_h + a\, g_a + n r'_h p_h + a r'_a p_a\right)$

Inserting (24) in (23) yields

(25) $p_h = B(C + D p_h)$

and thus

(26) $p_h = (1 - BD)^{-1} BC$

B, C, D are defined by

(27) $B = \gamma_h^{-1} n\, r_h (\tilde{r}_h n + \tilde{r}_a a + \gamma_e)^{-1}$

(28) $C = n\, g_h + a\, g_a$

(29) $D = n r'_h + a r'_a A,$

or, with (22)

(30) $D = n r'_h + \frac{a^2}{n} \frac{\gamma_h}{\gamma_a} \frac{r_a r'_a}{r_h}$

The solution (26) requires

(31) $(1 - BD) > 0$

because $p_h > 0$, and $BC > 0$.

$(1 - BD) = 0$ means that no steady state solution exists in contrast to the assumption (14). In this case the complete time-dependent equations must be considered which lead to instability.

This instability is caused by a feedback loop SHI→PI→SHI inherent in our (time-dependent) equations. In the spirit of allometry, we are interested in the functional dependence of p_h (humans) and p_a(automata), i.e. human and automata activities, on parameters, in particular number n of citizens and number a of automata, and rate constants. In the conversion rates $r_h, r_a, \tilde{r}_h, \tilde{r}_a$ for each individual human or automaton, a number of preselection constraints enter, such as special kinds of transmission channels, their network structure, wired, wireless etc. For our analysis it will be sufficient to discuss the relative size, at least of some pairs of rate constants. The rate r_h determines the information transfer from SHI (city) to PI (humans), while $\tilde{r}_h$ the information transfer from PI (humans) to SHI (city). Their relative size can be determined as follows. We consider these processes ignoring all other processes. Then according to (1) and (13) we obtain for their sum.

(32) $\frac{d}{dt}(s+p) = (r_h - \tilde{r}_h)s$
$r_h - \tilde{r}_h = 0$ means that $s + p = const.$,

Here we can draw on our previous work where we have introduced the notions of deflation and inflation meaning here

(33) $\frac{d}{dt}(s+p) < 0$ (deflation), $\frac{d}{dt}(s+p) > 0$ inflation.
(34) In practice, we expect deflation, which means $r_h < \tilde{r}_h$ (at least on average). The same conclusion holds true for automata,
(35) $r_a < \tilde{r}_a$.

The size of $r_h(r_a)$ is a measure of the ability of humans (automata) to convert SHI into PI, i.e. to recognize signals and to convert them into action. At present, surely $r_h \gg r_a$, but the concept of a "truly" smart city might imply $r_a > r_h$. In view of the big progress made in computer linguistics, the experimental determination of the transfer rates r per human (or automaton) where r may depend on the finite range of included topics is in our opinion possible. The r s we use in our eqs. are average values over many humans (automata) relevant for city life (including "import", "export"). Little can be said about the decay rates γ, but surely γ_e will increase in the course of time because of spams etc. When dealing with smart cities we must be careful when comparing the rates $r_h, \tilde{r}_h$ with $r_a, \tilde{r}_a$, because our use of the word automaton may comprise a wide range of interpretations—from smart household devices over robots till large computer centers. So when we consider special cases, we will have to take these distinctions into account. In particular, when computer centers play an important role, we have to compare the combinations $r_h n, \tilde{r}_h n$ with $r_a a, \tilde{r}_a a$. Finally we have to discuss the inclusion of export and import of material and immaterial goods. The effect of import (or "input") can be taken care of by extending the interpretation of D_h (6), D_a (7). In (6), we may replace the spontaneous generation rate g_h by

(36) $g_h = g_{hc} + g_{hi}$
where g_{hc} is just the former spontaneous rate, whereas g_{hi} is the generation rate of SHI induced by imports. Similarly, we may proceed with (7), by making the replacement.
(37) $g_a = g_{ac} + g_{ai}$
Note that, at least at the present state of AI, we may put
(38) $g_{ac} \approx 0$,
because we assume that AI-devices can hardly produce truly novel ideas. The rate of export, E, can be expressed by $E = r_{h\,e} p_h + r_{a\,e} p_a$ with the rate constants $r_{h\,e}, r_{a,e}$. In our approach, the SHI, PI dynamics is not directly influenced by E, but indirectly via the loss rates γ_h, γ_a. While clearly our model refers to a whole city and all its activities, a number of similar models can be established to deal with specific issues. Just to mention a typical example: humans observe ("measure") the walking speed of other humans (SHI) and respond by adjusting their own speed (PI) taking into account other relevant features of the city, in particular its size (e.g. Haken and Portugali 2016).

14.4 Special Cases Giving More Insight

In this section we discuss the dependence of p_h on the various parameters, particularly on n and a but also on the rate constants r. The special cases we treat are characterized either by the dominance of human activities over those of automata (case h), or vice versa (case a). Our starting point is (26) with its quantities B, C, D (27–30). The just mentioned cases (h, a) can best be distinguished by means of D (30). Multiplying (30) by $n\, r_h \gamma_a$ we obtain on the r.h.s. a sum of

(39) $(h)\ n^2 r_h r'_h/\gamma_h$ that refers to human "attributes" and

(40) $(a)\ a^2 r_a r'_a/\gamma_a$ that refers to automata "attributes"

According to $(h) \gg (a)$ or $(h) \ll (a)$, we derive simplified expression for p_h. First we assume

$$(h) \gg (a)$$

D (30) reduces to $D = nr'_h$, while B (27) reduces (under the assumptions $\tilde{r}_h n >> \tilde{r}_a a + \gamma_e$ and $\tilde{r}_h \approx r_h$) to

(41) $B = \gamma_h^{-1}$

so that

(42) $BD \approx n \frac{r'_h}{\gamma_h}$

Putting (26), (41), and (28) together, we obtain for the "load" per person

(43) $p_h/n \approx \left(1 - n\frac{r'_h}{\gamma_h}\right)^{-1} \gamma_h^{-1} \left(g_h + \frac{a}{n} g_a\right)$

where the impact of the term $\frac{a}{n} g_a$ is, in the frame of our approach, negligible. Most remarkable is the factor $\left(1 - n\frac{r'_h}{\gamma_h}\right)^{-1}$ which causes an enhancement of the spontaneous "production rate" g_h. This factor may lead to an instability, when $\left(1 - n\frac{r'_h}{\gamma_h}\right) = 0$ which may be possible for a sufficiently large population. In this case, no steady state of PI (and SHI) is possible. (See below). In the context of smart city, the case of the dominance of automata (a) is still more interesting, where $(h) \ll (a)$.

We assume $\tilde{r}_a \approx r_a$ and obtain.

(44) $B = \gamma_h^{-1} n\, r_h (r_a a)^{-1}$

(45) $D = a^2 r_a r'_a \gamma_h (n \gamma_a r_h)^{-1}$

so that

(46) $BD = a r'_a/\gamma_a$

Putting (26), (28), (45), and (46) together, we obtain for the personal load

(47) $p_h/n = \left(1 - \left(a\, r'_a\right)/\gamma_a\right)^{-1} \left(\frac{n\, r_h}{a\, r_a} \cdot \frac{g_h}{r_h} + \frac{g_a}{r_a}\frac{r_h}{\gamma_h}\right)$

This is our most important result. As the first factor reveals, "smartification", i.e. increasing $a\, r'_a/\gamma_a$ *enhances* the personal load and may even lead to instability. Because of the composition of $a\, r'_a/\gamma_a$, this conclusion holds for both many small AI devices and for few large computer centers. The first term in the second bracket in (47) is proportional to the number n of citizens and describes an increase of personal load proportional to the size of the population. Note, however, that we consider the case $n\,r_h \ll a\,r_a$, i.e. strong smartification. The last term in the second bracket represents the cooperation of automata and humans. Since we don't ascribe "creative" power to the automata, g_a depends solely on the *import* of goods, or external ideas. If the import g_{hi}, g_{ai} as well as personal initiative g_{hc} tend to zero, a dying city will result.

14.4.1 The Information Crisis

The basic *time-dependent* equations read.

(48) $\frac{dp_h}{dt} = n\,s\,r_h - \gamma_h p_h$

(49) $\frac{dp_a}{dt} = a\,s\,r_a - \gamma_a p_a$

(50) $\frac{ds}{dt} = n\,g_h + a\,g_a + n\,r'_h p_h + a\,r'_a p_a - \Gamma s$

where

(51) $\Gamma = \tilde{r}_h n + \tilde{r}_a a + \gamma_e$

To study stability, we consider the homogeneous eqs., i.e. $g_h = g_a = 0$, and make the hypothesis

(52) $p_h = e^{\lambda t} p_{h0}, \quad p_a = e^{\lambda t} p_{a0}, \quad s = e^{\lambda t} s_0$

Inserting (52) into (48)–(50) leads to the eigenvalue equation.

(53) $(\lambda + \Gamma)(\lambda + \gamma_h)(\lambda + \gamma_a) - nr'_h(\lambda + \gamma_a)nr_h - ar'_a(\lambda + \gamma_h)ar_a = 0$

For our purpose it suffices to consider the case where the automata are dominant.

(54) $a^2 r'_a r_a >> n^2 r'_h r_h$

Then (53) can be reduced to $\lambda = -\gamma_h$ or

(55) $(\lambda + \Gamma)(\lambda + \gamma_a) - a^2 r'_a r_a = 0$

The solution to (55) reads

(56) $\lambda_{+,-} = \frac{\Gamma + \gamma_a}{2} \pm \frac{1}{2}\left((\Gamma - \gamma_a)^2 + 4a^2 r'_a r_a\right)^{1/2}$

Instability leading to exponential growth occurs if $\lambda_+ > 0$. At the instability, $\lambda_+ = 0$, we find

(57) $\Gamma\gamma_a - a^2 r'_a r_a = 0, \quad \Gamma \approx \tilde{r}_a a$

which under the same approximations as above, leads to the previous instability condition

(58) $1 - \frac{ar'_a}{\gamma_a} = 0$

The exponential increase of SHI will lead to a communication breakdown so that this instability must be avoided. This can be achieved by technical innovations lowering the production rate of SHI e.g. by reducing the transformation rate r'_a by many little improvements or a global breakthrough in AI. A "typical" approach will be increasing the costs of SHI—use (including taxation).

14.5 Final Notes

We conclude this chapter with some notes on urban planning and design in smart cities: As we show in Chap. 15 that follows, the common view is to see urban planning and design as an external intervention in an otherwise spontaneous self-organized urban process. From this perspective, smart cities with their IOT (internet of things), equipped with sensors covering the whole city, data mining techniques that enable to dig into, and exploit, big data, provide urban planning and design stronger tools than ever to plan and design cities. Some would say, to transform the current self-organized, somewhat chaotic, cities into centrally organized cities. And what about the citizens? The answer is that part of the smart city machinery will be able to authentically identify and represent the citizens' views about the various plans and designs suggested by the professional planners/designers and implemented by the urban planning and design authorities. This approach, as can be seen, assumes a fundamental distinction between the professional planners and designers and the "planned" citizens of the city. This is the common view, as noted, and big companies (e.g. IBM) see here a potential future market. As we elaborate in Chap. 15, our view is different: The lesson from the study of cities as self organizing systems is that every urban agent is a planner at a certain scale and that in many cases, due to nonlinearities, the plan and design of a single person might be more dominant and influential than that of a whole team of professional planners. Examples are the stories of lofts, balconies etc. (cf. Chaps. 4 and 10 above). This view is further supported by cognitive science findings regarding humans' cognitive chronestetic capabilities for mental time travel and the implied phenomena of cognitive planning and prospective memory (cf. Chap. 15; Portugali 2016a). From this perspective, urban dynamics is seen as on-going self-organized and organizing interaction between the city's many agents/planners each with its specific local-, mezzo- or global-scale plan. The challenge of smart cities with their IOT would be to develop means to foster this process. In fact, Alexander's et al. (1977) *Pattern Language* and the Self-Planned City (Portugali 2011, Chap. 16) are steps toward this aim. There is an additional thought in particular brought forward by Haken (unpublished). The ongoing development and use of AI/IT leads us to reconsider problems in mathematical terms, because quite clearly AI is based on mathematics. According to mathematics there are several classes of problems.

(1) Under given conditions, a problem (e.g. in decision making) cannot be solved at all. This is not a purely academic issue, but may have consequences when AI is applied to real life problems (possibly even to traffic steering). Thus we must not expect miracles from AI.
(2) In many cases a problem possesses several solutions (cf. Chaps. 5 and 6). In sociological context this may lead to a "conflict" situation. Two simple examples may illustrate the case:

traffic

(a) left hand drive or

(b) right hand drive?

Both regulations are possible, but exclude each other. This conflict can be solved only collectively via governement decisions or direct votes.

name giving to married people

(a) regulated by state law
(b) left to couples

In this case also, the "conflict" is solved collectively. This example may seem far fetched, but in fact, conflict situtations are ubiquitous in cities and can hardly be solved by automata.

(3) In a number of cases, there exists a *unique* solution. In such a case, it doesn't matter, whether the solution is found by humans or automata.
(4) In multi-component/multi-agent/complex systems the processes are of a mixed deterministic/stochastic nature. This requires the study of scenarios (realizations) by a combination of human imagination and insights (see above) and computing power.

A city may be considered as laboratory for innovations for a better quality of life (cf. e.g. Batty et al. 2012) in which the decisions relevant to human welfare must be left to the citizens. In this chapter, aside from some general remarks, we focussed our discussion on the role of AI and information dynamics in cities. We believe that innovations by AI will play an ever increasing role, especially when dealing with the "information crises".

As the name indicates, AI directs attention to the role of artifacts, the production of which forms one of the basic capabilities of humans. That is, the production of objects that in one way or the other replace the natural capabilities of humans by artificial ones. Thus, some of the early stone tools (e.g. flint knives) replaced the natural human teeth and fingernails as cutting devices. As noted in the introduction, the emergence of cities some 5500 years ago was associated with the invention of writing—among the "smartest" inventions of society—which has partly replaced human memory: A person or society can write their story or thoughts on a stone (or papyrus or tablet or paper or computer) and need not keep it in memory. Already in antiquity this situation entailed a dilemma about the relations between the artifact and the natural human capability. In Plato there is an interesting dialogue between Socrates and Phaedrus in which Socrates expresses his concern about writing:

> In fact, it will introduce forgetfulness into the soul of those who learn it: they will not practice using their memory because they will put their trust in writing, which is external and depends on signs that belong to others, instead of trying to remember from the inside, completely on their own. You have not discovered a portion for remembering, but for reminding; you provide your students with the appearance of wisdom, not with its reality. Your invention will enable them to hear many things without being properly taught, and they will imagine that they have come to know much while for the most part they will know nothing. And they will be difficult to get along with, since they will merely appear to be wise instead of really being so." (Plato. c. 399–347 BCE. "Phaedrus." pp. 551–552 in *Complete Works*, edited by J. M. Cooper. Indianapolis IN: Hackett.)

This concern about innovative artifacts accompanies society for many years and is relevant today with the smart artifacts and cities. Looking back at history we can see that writing was not associated with the deterioration of memory: Rather it enabled the externalization and thus the extension of memory—a new form of division of labor between the artificial and natural memory (see SIRN in this respect). As in division of labor in general, so in the case of writing, the challenge was to find a steady state that maximizes the relative advantage of the human memory and that of the artificial memory. The same applies here: the challenge facing smart cities is to identify a steady state that maximizes the relative advantage of the human sensorium and intelligence, and that of the artificial ones.

References

Alexander, C., Ishikawa, S., & Silvestein, M. (1977). *A Pattern language*. New York: Oxford University Press.

Barabási, A.-L. (2016). *Network science*. Cambridge: Cambridge University Press.

Batty, M., Axhausen, K.W., Gianotti, F., Pozdnoukhov, A., Bazzani, A., Wachowicz, M., Ouzounis, G., & Portugali, Y. (2012). Smart cities of the future. *European Physics Journal Special Topics, 214*, 481–518. https://doi.org/10.1140/epjst/e2012-01703-3.

Bauernhansl, T. (2015). Arena 2036—The fourth industrial revolution in the automotive industry. In C. Glanz et al. (Eds.), *15th Stuttgart international symposium. Automotive and engine technology* (Vol. 1). Springer Vieweg: Wiesbaden.

Bettencourt, L. M. A., Lobo, J., Helbing, D., Kühnert, C., & West, G. B. (2007). Growth, innovation, scaling, and the pace of life in cities. *Proceedings of National Academy Science U.S.A., 104*(17), 7301–7306. https://doi.org/10.1073/pnas.0610172104.

Bonnefon, J.-F., Shariff, A., & Rahwan, I. (2016). The social dilemma of autonomous vehicles. *Science, 352*(6293), 1573–1576.

Bretagnolle, A., Daudé, E., & Pumain, D. (2006). From theory to modelling: urban systems as complex systems. *Cybergeo: European Journal of Geography. Dossiers 2005–2002, 335*. 13th European colloquium on theoretical and quantitative geography, Lucca, 8–11 September 2003.

Feder-Levy, E., & Blumenfeld-Liebertal, E., & Portugali, J. (2016). The well-informed city: A decentralized, bottom-up model for a smart city service using information and selforganization. IEEE International Smart Cities Conference (ISC2), Trento (pp. 1–4). https://doi.org/10.1109/ISC2.2016.7580767

Haken, H., & Portugali, J. (2016). Information and Selforganization. A unifying approach and applications. *Entropy, 18*, 197. https://doi.org/10.3390/e18060197.

Le Cun, Y., Bengio, Y., & Hinton, G. (2015). Deep learning. *Nature, 521*, 436.

Neuman, J. V., & Morgenstern, O. (1944). *Theory of games and economic behavior*. Princeton: Princeton University Press.

Nowak, M. A. (2006). *Evolutionary dynamics*. Cambridge, Mass., Belknap.

Portugali, J. (2000). *Self-organization and the city*. Berlin/Heidelberg/New York: Springer.

Portugali, J. (2011). *Complexity, cognition and the city*. Berlin/Heidelberg/New York: Springer.

Portugali, J. (2012). Complexity theories of cities: achievements, criticism and potentials. In J. Portugali et al. (Eds.), *Complexity theories of cities have come of age* (pp. 47–62). Heidelberg, Berlin: Springer.

Portugali, J. (2016). *Interview in Lisa Kremer: What's the buzz about smart cities?* Tel Aviv University.

Roscia, M., Longo, M., & Lazaroiu, G. C. (2013). Smart city by multi-agent systems. In *Proceedings of ICRERA 2013 IEEE international conference on renewable energy research and applications* (pp. 371–376).

Schwab, K. (2016). *The fourth industrial revolution*. Kindle Edition. Switzerland: World Economic Forum.

West, J. H., & Brown, G. B. (2005). The origin of allometric scaling laws in biology from genomes to ecosystems: Towards a quantitative unifying theory of biological structure and organization. *Journal of Experimental Biology, 208*, 1575–1592. https://doi.org/10.1242/jeb.01589.

Chapter 15
Cognitive Planning and Professional Planning

15.1 Introduction

Cities, as noted in Chap. 2, are hybrid complex systems composed, on the one hand, of artifacts which are essentially simple systems, while on the other, of human agents which are by their nature complex systems. And, it is the human agents that by means of their behavior and action make cities complex. In order to understand the way human agents act and behave in cities we have to consult the cognitive science—a domain of research that emerged in the mid 1950 as the science of mind, brain, body and their interrelations (Gardner 1987). The consultation with the cognitive science has led to several insights of which one stands at the core of the present discussion, namely, that there are two forms of planning associated with the dynamics of cities: *cognitive planning* as a basic cognitive capability of humans and thus of each individual urban agent, and professional/institutional planning, referring to governmental attempts to regulate the dynamic of cities. This chapter shows that these two forms of planning are interrelated via the processes of SIRN (*synergetic inter-representation networks*), IA (*information adaptation*) and their conjunction (SIRNIA). The chapter further shows that cognitive and institutional forms of planning are interwoven with each other in a kind of circular causality—a process that is central to the dynamics of cities as complex adaptive systems (Chap. 11).

The discussion below starts (Sect. 15.2) by introducing the notion of cognitive planning and its cognitive origin. Section 15.3 presents the SIRNIA view on urban planning. Section 15.4 that describes the dynamics of cities as complex systems, paves the way to Sect. 15.5 that illustrates how cognitive and institutional forms of planning are interwoven with each other, and how this interplay between them participates in the overall dynamics of cities as complex systems.

H. Haken and J. Portugali, *Synergetic Cities: Information, Steady State and Phase Transition*, Springer Series in Synergetics,
https://doi.org/10.1007/978-3-030-63457-5_15

15.2 Planning Through the Looking Glass

15.2.1 Chronesthesia and the City

> I don't understand you,' said Alice. 'It's dreadfully confusing!' 'That's the effect of living backwards,' the Queen said kindly: 'it always makes one a little giddy at first – 'Living backwards!' Alice repeated in great astonishment. 'I never heard of such a thing!' ' – but there's one great advantage in it, that one's memory works both ways.' 'I'm sure *mine* only works one way.' Alice remarked. 'I can't remember things before they happen.' 'It's a poor sort of memory that only works backwards,' the Queen remarked. [Conversation between Alice and the White Queen in *Through the Looking Glass* by Lewis Carrol]

While counter-intuitive, we now know that the White Queen is right; not only in Lewis Carrol's imaginative world, but in our everyday life. Originally hypothesized by Tulving (1983) with respect to episodic memory, the notion of *Chronesthesia*, or *mental time travel* (*MTT*) as it is also termed, refers to the brain's ability to think about—"mentally travel" to— the past, present, and future. Tulving's hypothesis was recently supported by neurological studies indicating that certain brain regions "were activated differently when the subjects thought about the past and future compared with the present. *Notably, brain activity was very similar for thinking about all of the non-present times (the imagined past, real past, and imagined future)*" (Nyberg et al. 2010). Schacter et al. (2008) have termed these processes "the prospective brain, whose primary function is to use past experiences to anticipate future events."

MTT provides a common denominator to several cognitive processes among them cognitive processes that support episodic simulation of future events (Schachter et al. ibid.) and *prospective memory*, referring to human ability to remember to perform an intended or planned action (McDaniel and Einstein 2007; Haken and Portugali 2005).

Following the above studies, it has been suggested (Portugali 2016a, b), firstly, that cognitive planning and design are direct manifestations of humans' chronesthetic memory, with the implication that humans are natural planners and designers. Secondly, that not only humans have the ability to mentally travel in time, but that they cannot avoid doing so. This is evident from the findings that "mind wandering", or, "stimulus-independent thought" is the brain's default mode of operation (Raichle et al. 2001; Buckner et al. 2008); and, "unlike other animals," humans spend about half of their waking hours "thinking about what is not going on around them, contemplating events that happened in the past, might happen in the future, or will never happen at all" (Killingsworth and Gilbert 2010).

It has thirdly been suggested, that urban agents are of no exception—they too are natural planners and designers. 'Natural' in the sense that planning and design are innate cognitive properties of humans, or more specifically, of human *agency*. As recently shown by Kelso (2016) in his "On the self-organizing origins of agency", human agency, that is, the capability for an "action toward an end", starts at very early age when the spontaneous movements of a baby "cause the world to change". The suggestion in this chapter is that, as a consequence of the above, not only that humans as urban agents *can* plan and design, but that *urban agents cannot avoid to plan or design.* [Obviously not all human action and behavior is planned and we thus need to distinguish between planned behaviors and un-planned behaviors]. Fourthly, that this natural human tendency to plan and design and the fact that we seem unable to not do it, is the cognitive generator behind several properties that make cities hybrid complex systems and thus differentiate cities from other forms of complex systems (Portugali ibid). To the above we here add that MTT is the cognitive generator of cognitive planning.

15.2.2 Cognitive Planning

Cognitive planning—the study of humans' cognitive capability to think ahead to the future and to act ahead toward the future, accompanies the cognitive sciences from their very origin. While there is a debate about the claim that planning is a property that separates humans from the rest of animals, there is no doubt that planning is specifically characteristic to humans. In their *Plans and the Structure of Behavior*, Miller et al. (1960) suggest seeing a 'plan' as analogous to a computer program and 'planning' as hierarchical problem-solving operation that guides action. To this view Hayes-Roth and Hayes-Roth (1979) have added the notion of *opportunistic planning* referring to cases where the person/planner responds to opportunities as they come, while Friedman and Scholnick (1987) and Das et al. (1996) contrasted opportunistic planning with 'global' and *hierarchical planning* (Ormerod 2005). Davies (2005) further added a distinction between *well-defined planning* where the required information is available at the start of the planning process versus *ill-defined planning* that commences with only part of the required information (Ormerod 2005). Ill-defined decision and planning situations are associated, on the one hand with Simon's (1957) *bounded rationality* (Newell et al. 1958) while on the other, with Tversky and Kahneman's (1974, 1981) set of five decision *heuristics* that people tend to employ in situations of high uncertainty in decision making and planning.

Looked upon from the perspective of cities as complex systems, the notion of cognitive planning implies and confirms the view, firstly, that every urban agent is a cognitive planner. Secondly, that cities are characterized by two kinds of planning: cognitive and institutional. As shown in past studies (Portugali 2000) and below, due to nonlinearities that characterize cities as complex systems, in many cases the act of a single urban agent as a cognitive planner affects the urban landscape more than that of institutional planning.

In a recent empirical study Sela (2015, 2016) shows that urban agents as cognitive planners treat the products of institutional planning (i.e. long-term city plans) in an ambivalent way: On the one hand, they are skeptical (or realistic?) about the chances of long-term city plans to be implemented as originally planned, while on the other, as cognitive planners they nevertheless use them as future anchors in their decision making. This chapter goes one step further and sheds light at the interplay that takes place between cognitive and institutional planning and the implications thereof on the dynamics of cities. It shows, as noted, that this interplay is governed by the process of SIRNIA that is introduced next.

15.3 A SIRNIA View on Urban Planning

SIRN, IA and their conjunction SIRNIA have been described in some length in Chap. 4, (also in connection with Chaps. 7 and 10). In this section we shortly elaborate on their planning dimension which is illustrated graphically by Fig. 15.1. We'll describe first the SIRN component of SIRNIA and then its IA component.

15.3.1 The SIRN Component

In terms of planning, the SIRN component of SIRNIA refers to an urban agent as a cognitive planner that is subject to two flows of information (Fig. 15.1): Externally represented information that comes from the city and includes the products of institutional planning in the form of plans and planning policies, and internally represented information constructed in the cognitive planner's mind/brain. The interaction between the two, gives rise to external representations in the form of the agent's planned action and behavior in the city, and internally represented information in the form of feedback to the agent's mind/brain.

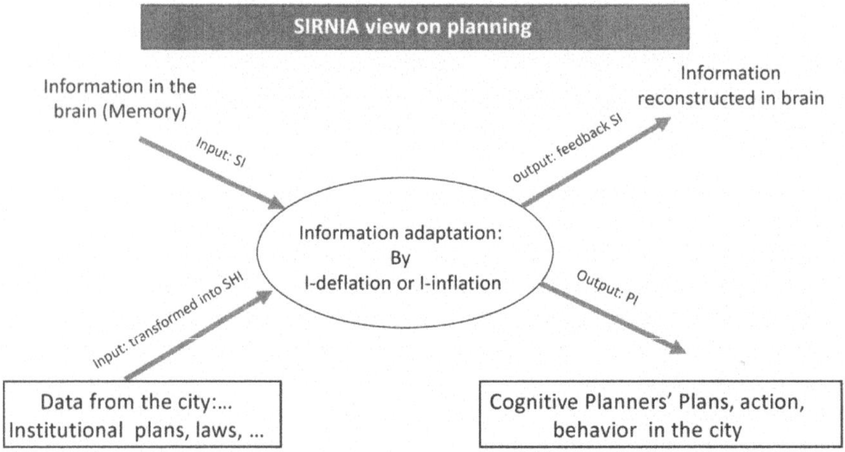

Fig. 15.1 SIRNIA view on the play between cognitive planning and professional-institutional planning

In terms of the derived three SIRN sub-models (Chap. 4, Fig. 4.2), we can say that the *intrapersonal* (Fig. 4.2, *top*), refers to a solitary agent, e.g. a planner/designer at work; the *interpersonal-sequential* (Fig. 4.2, *middle*), refers to a sequential dynamics of several agents, e.g. the space-time diffusion of a certain planning solution; and the *interpersonal collective* (Fig. 4.2, *bottom*) models the simultaneous interaction among many agents. At a small scale, the third sub-model might refer to a planning team engaged in a planning process (Portugali and Alfasi 2008); at a larger scale this third SIRN sub-model is also a model of urban dynamics (cf. Chap. 4).

15.3.2 The Information Adaptation Component

The IA component of SIRNIA (Fig. 15.1) refers to the interaction between the externally represented information that comes from the city and includes city plans, regulations and laws made by the city's professional planners, and the inrternally represented information constructed in the urban agent's memory. By means of the processes of information inflation or information deflation (cf. Chap. 4), this interaction gives rise to PI in the form of the plans, action and behavior the agent takes in the city, and SI that feeds back to the agent's memory.

The process, as we've seen in Chap. 4, is context dependent: when looking at a ground floor of a city building, officially defined by the institutional planning law as residential, one urban agent/cognitive planner might see an apartment—i.e. the PI of living, while a second urban agent/cognitive planner sees also (in his/her "creative" imagination) the potential to transform this apartment into an office, studio and/or a shop. When the latter agent/cognitive planner acts accordingly, thus violating the planning law, s/he creates a fluctuation/mutation in the city's steady state. In what follows we explore the way such an urban mutation affects the city and its dynamics.

15.4 Urban Dynamics

15.4.1 The Longue Durée of Cities

The *longue durée* of complex systems, cities included, is characterized by relatively long periods of steady state during which the system is structurally stable, followed by relatively short chaotic periods of strong fluctuations and instability that gives rise to a phase transition, morphogenesis, the emergence of a new structurally stable steady state and so on (Fig. 15.2). As we show in previous chapters, during the short chaotic periods local interactions between the parts give rise to the emergence of one/few order parameter(s) in a bottom–up manner, that then top–down dominate ("enslave") the system and the behavior of the parts and so on.

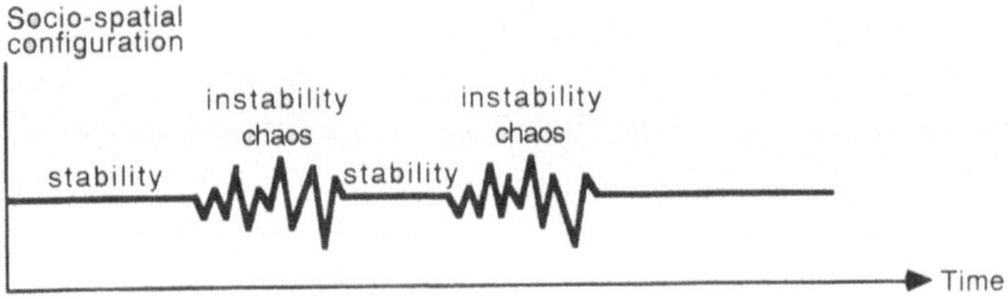

Fig. 15.2 The *longue durée* of complex systems

This typical systemic behavior can be explained by reference to the classic works of Malthus, Verhulst and Spencer. As is well recorded, Malthus' (1798) *An Essay on the Principle of Population*, entailed several reactions among them one by Verhulst (1838) who suggested a mathematical theory of populations and another by Spencer's (1852) "A Theory of Population". At the core of Malthus' view is an imbalance between linearly increasing means of subsistence versus exponentially growing population which must inevitably lead to a "population explosion." In contrast to Malthus, Verhulst emphasized the role of limit of resources—a concept that at a later stage was termed *environmental carrying capacity.* By means of the mathematical quadratic term, Verhulst demonstrated that the limit of environmental resources leads to the S-shape logistic growth curve (Fig. 15.3 *center*), that is, to the population adaptation. Spencer took Malthus theory and turned it upside-down: Human beings, he suggested, would not change their way of life unless they have to, and the Malthusian tension between the linearly increasing means of subsistence and exponentially growing population forces people and society to invent and change—it is one of the engines behind the processes of progress and innovation. Both Malthus and Spencer formulated their population theories verbally, but if we examine them in Verhulst's terms, they look as in Fig. 15.3 (*left* and *righ*).

Combining the "Verhulstian" notion of limit of resources as carrying capacity, with that of the synergetics' relations between control parameter and order parameter as discussed in previous chapters, we can say that once emerged, an order parameter has a certain carrying capacity. That is, it enables a balanced "freedom of choice" to

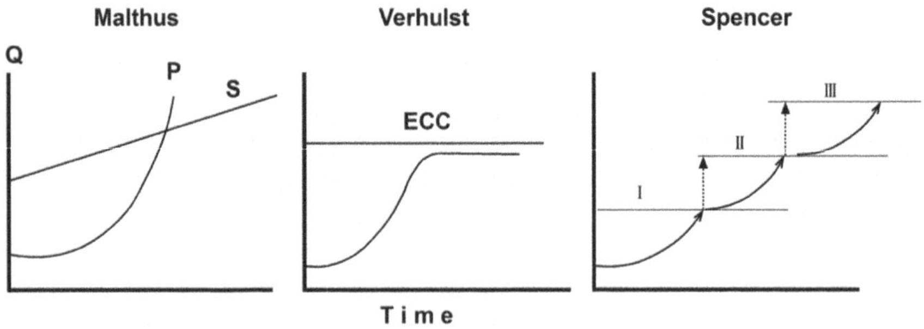

Fig. 15.3 *Left*—Malthus, *center*—Verhulst, *right*—Spencer. Q—quantity. P—Population. S—subsistence. EEC—environmental carrying capacity

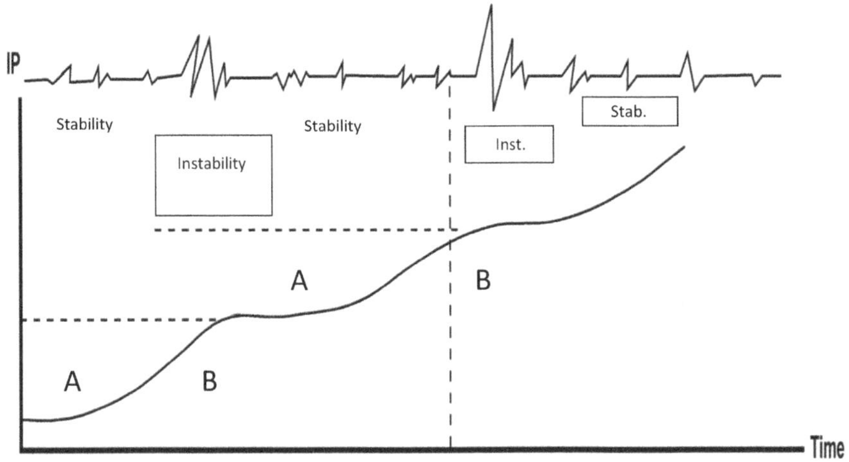

Fig. 15.4 Stability, instability and fluctuation in the *longue durée* of complex systems and of cities as complex systems. In periods A, the city's order parameter conveys high SHI with high freedom of choice; in periods B, little or no SHI/choice

the city's urban agents. As long as this carrying capacity is not exhausted, the order parameter is capable of "enslaving" the parts, a process of circular causality and structural reproductions occurs, the system is stable and resilient against internal and external perturbations. However, once the carrying capacity of the order parameter is for some internal or external reason exhausted, the system becomes unstable and loses its resilience, and thus vulnerable to internal or external fluctuations. The result is that the *longue durée of cities* evolves as a stepwise sequence of Verhulst's S-curves, driven by Spencerian innovations (Fig. 15.4). That is, long periods of steady state during which the city is resilient and thus enslaves local disturbances, followed by short chaotic periods during which the property of nonlinearity prevails (Fig. 15.4) so that a minor-scale local event might bottom–up give rise to a global phase transition and a large-scale global change.

15.4.2 Planning Order Parameters

According to synergetics' view on cities, the interaction between urban agents gives rise to one or a few order parameter(s) that once emerging, enslave(s), that is describe(s) and prescribe(s), the actions of the urban agents. As noted above (Chap. 11), a city as a complex system is typified by a hierarchy of systems and subsystems when each level in the hierarchy is dominated by its specific order parameter(s). A city can thus be described as a hierarchy of order parameters; for instance, a hierarchy of agents, neighborhoods, cities, metropolitan areas etc., each with its specific order parameter(s). In a similar way, one can speak on a hierarchy of order parameters related to various levels of planning such as metropolitan plans made by the metropolitan planning institutions and their many actors, city plans determined by the city planning authorities and their many actors, while at the bottom of the hierarchy local plans made by urban agents as cognitive planners. Note that the various institutional plans are of "Janus-face" character: from the top–down metropolitan plan's point of view the city plan is a bottom–up plan; from the top–down city plan the plans made by civil society organizations (NGOs etc.) are bottom–up plans while the local plans made by an individual urban agent is a bottom–up process. In contrast, the plans made by individual urban agents as cognitive planners are genuinely bottom–up. These bottom–up plans function as fluctuations/mutations in the overall behavior and evolution of a city. The above is, of course, an over-simplified description of planning which is not just plan-making activity but a process involving many factors, bodies and actors.

15.4.3 Fluctuations

As noted in previous chapters, complex systems are subject to on-going random fluctuations of various kinds, some of which are internally determined by the action and behavior of the parts of the system. As long as the system is in a stable steady state with a high level of carrying capacity, random fluctuations of the individual parts of the system have no significant effect on the overall/global evolution and behavior of the system. However, when the system enters an instable state (e.g. due to exhausted carrying capacity or any other internal or external reason) it loses its resilience and becomes vulnerable to such local fluctuations. At this stage, the system is specifically typified by non-linear effects in which a minor local perturbation (a "butterfly effect") might lead to a global phase transition. Well-studied examples come from synergetics (theory and experiment) studies on the nonequilibrium phase transition of lasers, from Weidlich's (1971) synergetic model of public opinion and perhaps less known, from studies on the role of mutations in evolution (Eigen 2013). A first application to cities is the discussion in Chap. 12 that applies the above properties to the case of urban scaling ("allometry"). In what follows the above aspects of fluctuations provide the basis for the interplay between cognitive and institutional forms of planning in the context of cities as complex systems.

15.5 The Interplay Between Cognitive and Institutional Planning

15.5.1 The City and Each of Its Urban Agents as Complex Adaptive Systems

Cities are dual complex systems in several respects of which two are relevant here: Firstly, in being hybrid complex systems as noted at the outset (cf. Chap. 2), and secondly, in the sense that the city as a whole is a Complex Adaptive System (CAS), and each of its parts—the urban agents—is a CAS too. From this follows two interrelated forms of adaptation in the dynamics of cities: Processes by which the urban agents adapt to the city's order parameter(s) by means of the synergetics' slaving principle (Chap. 11), and processes by which the city as a whole adapts to a bottom–up action initiated by urban agents.

As noted in the previous section, the city's planning law functions and can thus be treated as an order parameter. From this follows, on the one hand, processes by which the urban agents as cognitive planners adapt to (and thus enslaved by) the city's planning laws that function as the city's order parameter; while on the other, processes in which the city as a whole with its planning laws adapts to a bottom–up action initiated by urban agents as cognitive planners. From this further follows a distinction between two forms of actions initiated, and/or implemented, by urban agents as cognitive planners: plans/actions that conform with the city's order parameter-planning law and plans/actions that breach it.

The suggestion in this chapter is that plans/actions of individual cognitive agents that breach the planning law as order parameter, have the effect of, and can thus be treated as, fluctuations-mutations in the overall evolutionary dynamics of the system city. In line with the general dynamics of cities discussed above, as long as the fluctuations occur far from instability, i.e., in periods A of Fig. 15.4 when the order parameter affords freedom of choice, the city planning rules are resilient against such fluctuations and can and do enslave them when they occur. In such a situation, the urban agents as cognitive planners adapt to the city's planning order parameter. However, when as a consequence of some internal or external events the city's planning order parameter is low in its carrying capacity and thus allow no choices, the city enters a critical stage in which it operates close to instability, that is, "on the edge of chaos". At this situation, bottom–up small-scale fluctuations previously enslaved by the city's planning order parameter, can now lead the city into a phase transition and structural change. Such a situation is typical to cases where the city's planning order parameter adapts to local bottom–up plans initiated by urban agents as cognitive planners. But then, how exactly do bottom–up planning fluctuations emerge in the first place? The answer: by means the SIRNIA process introduced in Chap. 4 and above and further elaborated below with respect to planning.

15.5.2 Planning as Information Production

The city as an information production system was presented in Chap. 4 and Sect. 10.5. Here we add that institutional urban planning is a *par-excellence* information production process—producing information, on the one hand, about the current state (action possibilities) of the city, what e.g. land uses are allowed/prohibited in the various areas of the city, where to build what and so on. On the other hand, it informs the inhabitants about their city's "urban vision", the plans for the future, what will, or might, take place in the city, its possible future structure, and so on. This information which is the very content of the city's planning order parameter, can take the three forms discussed above, namely, quantitative Shannonian information (SHI), qualitative Semantic and/or Pragmatic information (SI and PI respectively).

Planning is an action oriented activity so that our main concern here is with SHI and PI or more specifically with PI determined SHI, that is, the quantity of SHI (action possibilities) conveyed by a certain planning order parameter. Here we can speak of the PI determined SHI of the city plan as a whole in which case this will be a measure of the order parameter's carrying capacity or freedom of choice, and also on the PI determined SHI of a given neighborhood or of a given building at a certain neighborhood. Following the logic of Haken and Portugali's (2003) study "the face of the city is its information" (cf. also Chap. 4, Sect. 4.2.2), the elementary principle is *urban diversity*: Low urban diversity refers to situations where the city's PI determined SHI is low, whereas high, to situations where the city's PI determined SHI is high. Applied to a city's land use plan for instance, a plan characterized by few land uses conveys little PI (i.e. choices of action for its citizens) and its PI determined SHI is low; whereas a city plan characterized by multiplicity of land uses conveys high PI determined SHI (i.e. choice of locational actions).

The same applies to a neighborhood or even a single building: If according to the city plan they are assigned mixed uses they convey high SHI, if single use, little. In line with the above, a plan conveying high PI/SHI has high carrying capacity/freedom of choice and is on the whole relatively stable and resilient against fluctuations, while plans conveying low SHI has low and is relatively unstable and vulnerable to local fluctuations.

So far the discussion referred to the information produced and conveyed by the institutional planning of a city. But there is another point of view—that of the urban agent as cognitive planner. For the latter, the information conveyed by a city, neighborhood or a single building is more akin to Gibson's (1979) notion of *affordances*. Namely, the action possibilities (PI) afforded by a given building, a neighborhood or a whole city. The latter information is determined, on the one hand, by the prevailing planning law, while on the other, by the very structure of the building, neighborhood or the city. For example, an apartment can be used for living (for which it was originally designed) but with minor modifications it can be transformed to an office, shop, studio etc. As the case studies in Chap. 4, Sect. 4.4 illustrate, this tension between the information produced by the institutional planning authorities versus the information conveyed by the very structure of an urban structure as perceived by cognitive planners, is one of the major engines of the dynamics of cities.

15.5.3 A Threefold Conjunction

The driving force behind the above two cases (and others that are discussed in Chap. 4, Sect. 4.4), is a threefold conjunction between the process of IA, the above noted process of planning as information production, and Weaver's view in his collaborative book with Shannon (Shannon and Weaver 1949, 9), that "...information is a measure of one's freedom of choice ..." From this conjunctive perspective, every urban element conveys to the city's cognitive planners/agents two forms of information: The PI determined SHI produced by the city's plan, versus the PI determined SHI afforded by the structure of the urban element. For example, according to the institutional plan, the buildings in NY and the balconies in TA had one optional use only: warehouse and open balcony (cf. Chaps. 4 and 10). There is no freedom of choice here and accordingly the PI determined SHI in each case was 0 bits, that is: $I = \log_2 1 = 0$. On the other hand, from the perspective of the cognitive planner, the warehouse and the balcony afford more choices and thus information: For the creative cognitive planner back in the 1960s NYC, the warehouse, for instance, conveyed three possible uses: apartment, studio and shop, that is, $I = \log_2 3 \approx 1.5$ bits of SHI and so on.

15.5.4 Urban Cognitive Planners as Promoters and Preventors

Central to the above processes is a chronic tension between institutional and cognitive planning forms. And yet, at least on the face of it, there should be no tension between them: Urban agents are expected to observe the law and act according to the information produced by the city planning law and plans. Yet reality is more complex than that. Firstly, laws in general and planning laws included, are never fully unequivocal; rather, more often than not, they can be interpreted in more than one way. In society at large, one of the major functions of the judicial system is thus to interpret the law. In a similar way, a major role of the various urban planning committees is to interpret the city's plans and its planning laws. This process of interpretation and reinterpretation of the planning law provided the foundation for the proposal to re-structure the planning process in such a way that it will enable a self-organized, "Self-planned city" (see Portugali 2011, Chap. 16 and full bibliography there).

Secondly, people—urban agents as cognitive planners in our case—differ in the way they tend to take decisions in face of a given planning law. Some accept the law as it literally is and take decisions accordingly, while others, in order to achieve their goals, try to interpret the planning law in a creative way or even to take a risk and breach it. Where do such behavioral differences come from? The suggestion here is that an answer can be offered by reference to the *urban regulatory focus theory* we discuss in Chap. 13. As noted there, urban agents' goal-directed behavior is regulated by two motivational systems, *promotion* and *prevention*: some urban agents are more promotion-oriented while others are more prevention-oriented. Promotion oriented agents tend to take risks and "check borders", whereas prevention motivated urban agents tend to avoid unnecessary risk.

The suggestion (or rather working hypothesis) here is that the promotion-prevention tendencies of urban agents participate in the dynamics of cities in the following way: prevention oriented cognitive planners tend to conform with the prevailing institutional planning order parameter and accept the planning law as it literally is, while promotion oriented cognitive urban planners, in order to achieve their goals, try to interpret it in an innovative way and from time to time even to breach the planning law. These promotion-derived innovations and breachings create fluctuations in the evolving urban system.

Cities as complex systems are subject to on-going random fluctuations-mutations of various kinds (Sect. 4.3, above), some of which are the result of the decisions and actions of the promotion oriented urban cognitive planners. As implied by the above discussion and case studies, such fluctuations are likely to happen when there is a wide information gap (measured by PI determined SHI) between the information conveyed by the city institutional plan to that afforded by the very structure of the city or elements in it. As further discussed above, the effect of such fluctuations on the city is minor during periods in which the city is in a steady-state with high carrying capacity/freedom of choice, while specifically critical in periods of structural instability.

15.6 Conclusions

The prevalent view in the domain of complexity theories of cities is that planning is an external intervention in an otherwise complex system. The view suggested here is (Portugali 2000, 2011, 2016a, b) that planning is one among many urban agents that interact in the city. This follows, firstly, from the inner logic of complexity theories of cities, namely, due to the property of nonlinearity that typifies cities as complex systems. Secondly, due to the findings (Portugali 2016a, b) that humans, and by implication urban agents, are cognitive planners. This chapter takes this view one step further by theorizing about the interplay between cognitive planning and institutional planning and its role in the overall dynamics of cities. The discussion above is rather preliminary, however—a kind of research agenda that hopefully will lead to subsequent and more detailed studies. Ultimately, planning agencies are not external control parameters, but rather part of the system "city" underlining the laws of selforganization. Both urban agents as cognitive planners and planning agencies thus act in the context of, and are influenced by, the *Zeitgeist*—the spirit of the time, and by the *genius luci*—the spirit of place.

References

Buckner, R., Andrews-Hanna, J., & Schacter, D. (2008). The brain's default network: Anatomy, function and relevance to disease. *Ann NYAcad Sci, 1124,* 1–38.

Das, J. P., Kar, B. C., & Parrila, R. K. (1996). *Cognitive planning: the psychological basis of intelligent behavior*. New Delhi: Sage Publication.

Davies, S. P. (2005). Planning and problem solving in well-defined domains. In R. Morris & G. Ward (Eds.), *The cognitive psychology of planning* (pp. 35–51). Hove: Psychology Press.

Eigen, M. (2013). *From strange simplicity to complex familiarity: A treatise on matter, information, life and thought*. UK: Oxford University Press.

Friedman, S. L., & Scholnick, E. K. (1987). *Blueprints for thinking: The role of planning in cognitive development*. Cambridge: Cambridge University Press.

Gardner, H. (1987). *The mind's new science*. New York: Basic Books.

Gibson, J. J. (1979). *The ecological approach to visual perception*. Boston: Houghton-Mifflin.

Haken, H., & Portugali, J. (2003). The face of the city is its information. *Journal of Environment Psychology, 23,* 385–408.

Haken, H., & Portugali, J. (2005). A synergetic interpretation of cue-dependent prospective memory. *Cognitive Processing, 6,* 87–97.

Hayes-Roth, B., & Hayes-Roth, F. (1979). A cognitive model of planning. *Cognitive Science, 3,* 275–310.

Kelso, S. J. A. (2016). On the self-organizing origins of agency. *Trends in Cognitive Sciences, 20,* 490–499.

Killingsworth, D. T., & Gilbert, A. (2010). Wandering mind is an unhappy mind. *Science, 330*(6006), 932.

Malthus, T. R. (1798): *An essay on the principle of population, as it affects the future improvement of society with remarks on the speculations of Mr. Godwin, M. Condorcet, and other writers.* Anonymously Published.

McDaniel, M. A., & Einstein, G. O. (2007). *Prospective memory: An overview and synthesis of an emerging field*. London: Sage.

Miller, G. A., Galanter, E. H., & Pribram, K. H. (1960). *Plans and the structure of behavior*. New York: Holt Rinehart & Winston.

Newell, A., Shaw, J. C., & Simon, H. A. (1958). Elements of a theory of human problem solving. *Psychological Review, 65,* 151–166.

Nyberg, L., Habib, R., Kim, S. A. N., Levine, B., & Tulving, E. (2010). Consciousness of subjective time in the brain. *Proceedings of National Academy of Sciences, 107*(51), 22356–22359.

Ormerod, T. C. (2005). Planning and ill-defined problems. In R. Morris & G. Ward (Eds.), *The cognitive psychology of planning* (pp. 53–70). Hove: Psychology Press.

Portugali, J. (2000). *Self-organization and the city*. Berlin/Heidelberg/New York: Springer.

Portugali, J. (2011). *Complexity, cognition and the city*. Berlin/Heidelberg/New York: Springer.

Portugali, J. (2016). *Interview in Lisa Kremer: What's the buzz about smart cities?*. Tel Aviv University.

Portugali, J. (2016b). What makes cities complex? In J. Portugali & E. Stolk (Eds.), *Complexity, cognition, urban planning and design*. Berlin/Heidelberg/New York: Springer.

Portugali, J., & Alfasi, N. (2008). An approach to planning discourse analysis. *Urban Studies, 45*(2), 251–272.

Raichle, M. E., MacLeod, A. M., Snyder, A. Z., Powers, W. J., Gusnard, D. A., & Shulman, G. L. (2001). A default mode of brain function. *Proceedings of National Academy Science USA, 98*(2), 676–682.

Schacter, D., Addis, D. R., & Buckner, R. (2008). Episodic simulation of future events: concepts, data, and applications. *Annals N.Y. Academy Science, 1124*, 39–60.

Sela, R. (2015). Cognitive aspects of planning and their implications for urban planning. PhD dissertation, TAU.

Sela, R. (2016). Global scale predictions of cities in urban and in cognitive planning. In J. Portugali & E. Stolk (Eds.), *Complexity, cognition, urban planning and design.* Berlin/Heidelberg/New York: Springer.

Shannon, C. E., & Weaver, W. (1949). *The mathematical theory of communication.* Illinois: University of Illinois Press.

Simon, H. A. (1957). A behavioral model of rational choice. In H. A. Simon (Ed.), *Models of man, social and rational: Mathematical essays on rational human behavior in a social setting.* New York: Wiley.

Spencer, H. (1852). A theory of population Westminster Review April. In R. L. Carneiro (Eds.) (1967), *The evolution of society: Selections from Herbert Spencer's Principles of sociology.* Chicago: University of Chicago Press.

Tulving, E. (1983). *Elements of episodic memory.* Oxford: Clarendon Press.

Tversky, A., & Kahneman, D. (1974). Judgment under uncertainty: heuristics and biases. *Science, 185,* 1124–1131.

Tversky, A., & Kahneman, D. (1981). The framing of decision and psychology of choice. *Science, 211,* 453–458.

Verhulst, P. F. (1838). Notice sur la loi que la population suit dans son accroissement. *Correspondence Mathematical and Physics Publication Par A. Quetelet, X,* 113–121.

Weidlich, W. (1971). The statistical description of polarization phenomena in society. *British Journal of Mathematical and Statistical Psychology, 24,* 251–266.

Chapter 16
Concluding Notes

The notion 'synergeticcities' that forms the core of this book evolved gradually in the last decades out of the collaboration between the two of us. In a way, the various chapters in the three parts of the book follow this evolution. Thus, Part I of the book starts with the founding formulation of Synergetics by means of the LASER paradigm and discusses its implications to cities (Chap. 3); it continues with SIRNIA (Chap. 4), that was inspired by the synergeticpattern recognition paradigm, as well as by the need to add cognition to our understanding of citiesascomplex systems. Chaps. 5 and 6 elaborate the 1st and 2nd Foundations of Synergetics and their urban implications, while Chap. 7 suggests an innovative preliminary discussion on Friston's FEP, its relation to the Synergetics 2nd foundation as well as to our SIRNIA view on cities. Part II (Chaps. 8–11) suggests a synergeticperspective on steady statesand phase transitionsin cities—two aspects that are central tocomplex systems but for historical reasons have not been given sufficient attention in the CTC discourse. Finally, Part III explores the implications of the synergeticcities view to issues that stand at the center of current CTC discourse: urban allometry (Chaps. 12 and 13), smart cities (Chap. 14) and city planning (Chap. 15).

In concluding this book we want to suggest a few preliminary comments on a topic that we've mentioned only briefly in the book (Chap. 10, Sect. 10.3), but was always present while this book was written. We refer to the fact that major parts of this book were written during the period of time in which world society was in a crisis due to the COVID-19 pandemic. While it is too early to comment in full on this event, already at this stage it can be observed that the corona event exhibits many of the ingredients of acomplex system: abrupt change, non-linearity, unpredictability, uncertainty and more. So, a few notes related to our view on citiesascomplex systems might be in place.

In terms of our book, the whole event started as a minor fluctuationin the steady state of a city—a typical *butterfly effect*. This fluctuation took place at end 2019 when a *single person* (!!!) was infected with a virus from an animal at a *single location*—(probably) at the now world-famous Huanan seafood market of Wuhan,

H. Haken and J. Portugali, *Synergetic Cities: Information, Steady State and Phase Transition*, Springer Series in Synergetics,
https://doi.org/10.1007/978-3-030-63457-5_16

China. Within a few months the virus underwent a space-time diffusion process and turned from a minor local fluctuation into a global order parameter of world society, describing and prescribing the life and behavior of millions of people around the world.

As we show in our book, when a fluctuation occurs during a stable state, the system is resilient and enslaves it. However, when the system is in an instable state, a minor fluctuation can give rise to grand outcomes. Two interesting implications are associated with this dynamics—one, more theoretical, refers to a notion we term *latent instability*, while the second, more socio-political, refers to the 'crisis of democracy' that can be observed in recent years.

Latent instability. In retrospect we now know that the global system of cities was in a latent unstable and vulnerable state, "ripe" for an event such as the corona to erupt: a highly urbanized and geographically connected global society more than ever before in human history. Urbanization implies population densities, making it easy for the virus to spread from one person to the other, while high ground and air connectivity enable the fast spread of the virus from city to city all over the world.

Both urbanization and connectivity were always perceived as advantages that make global society and its cities economically, socially and culturally resilient. But the 2008 economic crisis was a hint that high "smart" connectivity has its drawbacks, while the 2011 *Fukushima* Daiichi nuclear disaster exposed the vulnerability of a highly dense urbanized society. Yet these lessons were not generalized to other domains. The dark sides of globalization, urbanism and connectivity were largely overlooked. Or more generally, what was overlooked is that each system has its weaknesses, that the very properties that make a system resilient in one domain, say, against economic fluctuations, might make it vulnerable to another kind of fluctuation. The implication is that a system's resilience is context or "fluctuation dependent".

Democracy. The motto that one hears time and again in the last decades is that society is becoming urban, that for the first time in human history more than 50% of human population live in cities, and so on. This is of course true, but: The 'but' is that this motto creates the wrong impression that by becoming more urban, world society is becoming also more uniform, that the old 'town-country antagonism' has gone as we all experience the same global urban reality. Yet this is not the case. The old town-country antagonism was replaced by a new spatial antagonism, this time between society in the big 'world cities' that form the hubs of the borderless global society and economy, vs. society in the local peripheral towns and cities confined by, and dependent on, national boundaries and governments. This tension took the form of a spectrum of positions at one pole of which is "non-liberal democracy while on the other, "non-democratic liberalism". This tension, as is well recorded, takes different forms in different countries: A partial list includes the Brexit in England, Trump's presidency in the USA, the '*Mouvement des gilets jaunes*' in France, and more. Some, like Mounk (2018) see this tension in terms of *The people versus Democracy,* some like Levitsky and Ziblatt (2018) claim that we are witnessing *How DemocraciesDie*, while Guilluy (2014) in his *La France périphérique* suggests seeing this tension in terms of the center versus the periphery. Democracy, that for half a century, since mid twentieth century, was marked by stability due to a delicate balance and complementary relations between liberalism and the rule of the 'Demos', is currently in a state of instability and strong fluctuations.

As noted throughout the book and above, when a system is in an instable state, a minor fluctuation can give rise to grand outcomes. What is and/or would be, the effect of the corona fluctuation on the crisis of democracy? Will it strengthen the side of the national(ist) demos? or that of global liberal society? Or will there be a bifurcation and both sides will go on?

And a final note. Our earth is a giant laboratory testing a variety of systems; so is also our socio-cultural-political globe. Various systems are being tested, ranging from fierce dictatorships through democracies to the primitive communism of hunters and gathers societies. But what is meant by "tested"? "Survival of the fittest",[1] coexistence in ecological niches that seemingly are more or less wiped out by globalization, or various kinds of symbiosis? What is the role of large companies especially in the fields of IT and AI? Concerning the "style" of government, it can be proved (Haken, unpublished) that adaptability times stability = constant.

Churchill once remarked that "many forms of Government have been tried, and will be tried in this world of sin and woe. No one pretends that democracy is perfect or all-wise. Indeed **it has been said**that democracy is the worst form of Government except for all those other forms that have been tried from time to time...." [(House of Commons, 11 November 1947) in *Churchill by Himself* 2008]. There is an often felt tension between the welfare of an individual, or of a social group and that of a society—is this tension an interpretation depending on ideology, or an outcome of the Miler's law aphorism, "Where you stand depends on where you sit."?. Some believe that they can control the course of history; is this an illusion? All in all, it seems that there will be no final outcome—nature's and society's tests will probably go on and on.

References

Guilluy, C. (2014). La France périphérique. Comment on a sacrifié les classes populaires, Paris, Flammarion.

Levitsky, S., & Ziblatt, D. (2018). *How Democracies Die*. Broadway Books N.Y.

Mounk, Y. (2018). *The people vs democracy: Why our freedom Is in danger and how to save it*. Harvard University Press

[1] A phrase coined by Herbert Spencer regarding Darwin's theory of evolution.

Bibliography

Bayes, T. (1763). An essay towards solving a problem in the doctrine of chances. In Prince (Ed.), *Philosophical transactions royal society* (Vol. 53, pp. 370–418). London.

Clifton, K. J., & Moura, F. (2017). Conceptual framework for understanding latent demand: Accounting for unrealized activities and travel. *Transportation Research Record Journal of the Transportation Research Board.,2668,* 78–83.

Floridi, L. (2015). *Semantic conceptions of information.* Stanford, CA, USA: Stanford Encyclopedia of Philosophy.

Holling, C. S. (1986). Resilience of ecosystems; local surprise and global change. In W. C. Clark, R. E. Munn (Eds.), Sustainable development of the biosphere (pp. 292–317). Cambridge: Cambridge University Press.

Jaynes, E. T. (1957). Information theory and statistical mechanics. *Physical Review,106,* 620–630.

Jaynes, E. T. (1957). Information theory and statistical mechanics II. *Physical Review,108,* 171–190.

Kadanoff, L. (1966). Scaling laws for Ising models near Tc. *Physics,2,* 263–272.

Lewis, C. *Through the looking glass.*

Maturana, H., & Varela, F. (1980). *Autopoiesis and cognition.* Dordrecht, Holland: Reidel.

Portugali, J., & Stolk, E. (Eds.). (2016). *Complexity cognition urban planning and design.* Heidelberg, Berlin: Springer.

Prigogine, I., & Stengers, I. (1997). *The end of certainty: Time, chaos and the new laws of nature.* New York: The Free Press.

Ross, G. M. (2015). Urban regulatory focus: Promotion and prevention are part of the identity of the city. PhD dissertation, TAU.

Stolk, E., & Portugali, J. (2016). A complexity-cognitive view on scale in urban design. In J. Portugali & E. Stolk (Eds.), *Complexity, cognition urban planning and design* (pp. 217–236). Heidelberg, Berlin: Springer.

Weidlich, W. (1999). From fast to slow processes in the evolution of urban and regional settlement structures: The role of population pressure. In J. Portugali (Ed.), *Population, environment and society on the verge of the 21st century. A special theme issue of discrete dynamics national society* (Vol. 3, pp. 137–147).

Zünd, D., & Bettencourt, L. M. A. (2019). Growth and urban development in prefecture-level China. *PLoS ONE,14*(9), e0221017. https://doi.org/10.1371/journal.pone.0221017.

H. Haken and J. Portugali, *Synergetic Cities: Information, Steady State and Phase Transition*, Springer Series in Synergetics,
https://doi.org/10.1007/978-3-030-63457-5

Subject Index

H. Haken and J. Portugali, *Synergetic Cities: Information, Steady State and Phase Transition*, Springer Series in Synergetics,
https://doi.org/10.1007/978-3-030-63457-5

The manufacturer's authorised representative in the EU is Springer Nature Customer Service Centre GmbH, Europaplatz 3, 69115 Heidelberg, Germany. If you have any concerns regarding our products, please contact ProductSafety@springernature.com

Printed and bound by CPI Group (UK) Ltd, Croydon, CR0 4YY
30/09/2026
02225694-0007